THE ECONOMICS OF NORTH SEA OIL TAXATION

Also by Danny Hann

GOVERNMENT AND NORTH SEA OIL

Also by Chris Rowland

THE ECONOMICS OF CHOICE BETWEEN ENERGY SOURCES (*co-editor*)

The Economics of North Sea Oil Taxation

Chris Rowland
Financial Analyst
BZW Securities,
London

and

Danny Hann
Economist with the
Central Electricity
Generating Board

St. Martin's Press New York

 For information, write:
Scholarly & Reference Division,
St. Martin's Press, Inc., 175 Fifth Avenue, New York, NY 10010

First published in the United States of America in 1987

Printed in Hong Kong

ISBN 0–312–23678–6

Library of Congress Cataloging-in-Publication Data
Rowland, C. (Chris)
The economics of North Sea oil taxation.
Bibliography: p.
Includes index.
1. Offshore oil industry—Taxation—Great Britain.
2. Offshore oil industry—North Sea. I. Hann, Danny.
II. Title
HD9560.8.G72R69 1987 336.2'783338232'0941 86–13729
ISBN 0–312–23678–6

To Barbs and Hanna

Contents

Acknowledgements

We would like to thank Professor Colin Robinson for providing both the opportunity to undertake this project and the guidance crucial to its completion, and our colleagues both past and present in the Surrey Energy Economics Centre of the Department of Economics, University of Surrey, for providing a stimulating academic environment by continually questioning our findings.

We would also like to acknowledge our parents, and Jon, Sue, Christine and Barbara for their support and encouragement.

CHRIS ROWLAND
DANNY HANN

1 Area for Concern

North Sea oil extraction represents one of the most profitable industrial ventures in the post-war UK economy. It is also an industry whose profits are subject to a plethora of tax regulations. Our book is concerned with the consequences of these regulations.

In May 1982, the Select Committee on Energy expressed some concern that the North Sea fiscal regime may act as a deterrent to future North Sea investment.[1] The Committee concluded:

> that the tax system, at its current level, complexity and frequency of change, has now passed the point at which its impact can be said to be 'broadly neutral' and that a substantial risk exists that development is being discouraged.[2]

Our study considers whether this conclusion is still valid or if the tax changes since 1982 have remedied any problems. The Committee found taxation to be a contentious area. Indeed, it was the area in their wide-ranging report where

> the strongest differences between the Department of Energy and the oil companies emerged. The Department maintained, in their initial evidence in May and June 1981, that the system was 'broadly neutral', did not inhibit the pace of development of the North Sea, and was not being used as a deliberate depletion control instrument . . . The companies claimed that the fiscal régime was indeed having an adverse effect on activity and that, whatever the government's intention might be, it was acting as a de facto tool of depletion policy.[3]

In response to the Select Committee, the Chancellor of the Exchequer and the Secretary of State for Energy outlined the Government's view:

> The Government's own analysis of the effects of the fiscal régime on North Sea development does however differ significantly from that made by the Committee.[4] On the basis of detailed analysis . . . and a critical appraisal of the industry's own representations, the Government believes that a wide range of potential developments in the UK sector of the North Sea remains attractive. It considers

> that the fiscal regime is fully compatible with a satisfactory flow of new developments.[5]

Moreover,

> The Government has devoted considerable effort and resources over the past few years to analysing a wide range of alternative tax systems . . . it does not believe that these alternatives would meet the various objectives involved more satifactorily than the structure it now [in March 1982] proposes.[6]

Nevertheless, large-scale changes in the oil taxation system were introduced by the Chancellor of the Exchequer in the March 1983 Budget. These changes were met with 'surprise and jubilation among the oil companies'.[7] We consider if government and industry contentment with the tax system is appropriate, and the government is justified in its view, that no alternative system would be an improvement.

One theme developed is that the effect of the North Sea oil tax system on the offshore oil extraction industry appears to be misunderstood. Attention has tended to focus on the aggregate level of taxation and on the average level of the tax take over the lifetime of a typical North Sea field. As a consequence the issue of taxation is seen as a matter of equity, where one side can be made better off only at the expense of the other side. Our analysis, however, shows that by altering the structure of taxation both sides can be made better off. By alleviating the fiscal deterrents on new development projects, offshore recoverable reserves will be increased. The distribution of the increased value may be a question of equity, but the failure to achieve the increase is a question of economic efficiency.

A dangerous level of complacency has now developed, especially since the 1983 Budget tax changes. There seems to be an inconsistency between industry's and government's perceptions of each other's requirements in the oil tax system. For instance, the industry believes taxation will not be made harsher and that the government would be willing to forgo short-term tax revenues. The government, on the other hand, does not believe damage may be done by increasing the burden of taxation on companies especially after capital expenditures have been sunk. Neither the government nor the industry seem to recognise the substantial deterrent still imposed by the oil tax regime on marginal development prospects.

Marginal oil–field projects would benefit the economy as they offer rates of return higher than those obtainable elsewhere by more than enough to compensate for higher risks. However, at the post-tax level these marginal fields are often not commercially attractive. This basic problem of the oil taxation system was not alleviated by the 1983 changes and the impact on the future development of oil fields remains. The scale of the problems both to the offshore industry and to the domestic economy is such that the tacit acceptance of the tax system (especially by the government which has a role in protecting the economic interests of the UK) and the failure to adopt an alternative system is of considerable concern.

Clearly, government contentment and its contention that no better alternative exists are connected. If negligible problems are created by the current tax system then improvements will be difficult. The thrust of this book is that development decisions are being deterred on a scale which is significantly greater than is perceived by government and a better alternative does exist.

Chapter 2 presents an analytical account of government North Sea policy and provides a background for the later, more detailed investigation of tax policy. Here we examine the targets of government North Sea policy and contrast them with the outcome of policy. We find that development decisions do seem to be impeded by several areas of government North Sea policy, but particularly so by the tax regime. Tax policy appears to be the most damaging aspect of North Sea oil policy while changes made to the system seem to have aggravated, not alleviated, the distorting impact of oil taxation on oil–field development.

The impact of tax policy on North Sea activity is analysed in Chapter 3 under five sub-headings; namely, regressivity, the return on capital, risk premia, taxing incentives and instability. Chapter 4 quantifies the impact of the existing tax system by exploring the extent of the damage caused by domestic oil taxation policy on the level of UK continental shelf (UKCS)recoverable reserves. Chapter 5 presents an alternative system of taxation which is not subject to the same defects and does not reduce recoverable reserves to the same extent. In Chapter 6 we suggest why a better, alternative tax policy has failed to be adopted and outline the factors which inhibit the abolition of the existing system.

Appendix A provides a description of the oil tax system and its evolution from 1975 to 1984.

2 Government Policy: Targets and Instruments

2.1 INTRODUCTION

This chapter attempts to discover desirable policy objectives for the North Sea. By predicting the outcome in an unregulated market and comparing it to the outcome where the maximum value added is obtained, we are able to ascertain desirable directions for government policy. We then examine stated government policy intentions and analyse, in the third section of the chapter, the effectiveness of the measures adopted to reach these intentions. Our concern is first, to see if government policy intentions are desirable, and second, to see if government policy functions in a way that improves on the unregulated market.

A term frequently used in the analysis is 'economic rent', and this is explained in the following section. Next an analytical representation of development and production decision process, referred to in subsequent sections, is presented.

Economic Rent

The economic worth of a licence to produce oil from a tract of the UKCS sea bed may be measured by the present value of the flow of future revenues from that tract's production less the present value of the flow of associated future costs, where the costs include monetary items such as equipment as well as non-monetary items such as exposure to risks. The difference between these two amounts, the net present value (NPV), is the economic rent of that tract. The economic rent may be positive, zero or negative. If it is positive, this implies the licensee is enjoying profits in excess of those necessary to induce the production of oil (known as pure profits). In the North Sea oil extraction industry, pure profits are not eliminated by competition. In non-North Sea activities, the existence of companies earning pure profits tends to attract new entrants into that activity and, in the long run, each firm will earn normal profits only, as either prices will tend to fall or costs increase. In the North Sea, because oil is non-renewable, pure profits cannot be eliminated in this fashion.

Providing pure profits are defined after *all* costs, their magnitude will not be relevant to economic decisions and hence they comprise a justifiable base for taxation.

A distinction should be drawn between pure profits and 'windfall profits' where the latter refers to gains (or losses) which were totally unanticipated. The taxation of windfall profits, since they are unanticipated, can in no way have an impact on economic decisions. However, if the gains (or losses) were even remotely anticipated then they should be viewed as a reward (or penalty) for bearing risks and hence must be excluded from pure profits.

The Development and Production Decision

Once a prospect has been drilled, companies will form plans about the way to develop and deplete any potential fields. Here we abstract from the planning process to show how companies simultaneously choose the level of commercially recoverable reserves and an output profile. The driving force behind this choice is assumed to be profit maximisation.

Conceptually there are two separate parts to field decision: first on the extent of capacity to install, and second on the level of operating the installed capacity. Consider the marginal revenues and costs in the second part of the decision. Depending on the company's relation to the industry, its marginal revenues may be independent of, or related to, the scale of output. In the international oil industry each company is effectively a price-taker and marginal revenues will be independent. For UK gas fields, the large international companies may believe that the price received for a marginal unit of production will affect prices on other units, especially while the international gas trade is poorly established, and marginal revenues may be inversely related to output. Marginal costs will probably be positively related to output. As production in a given period expands, more and more resources will be needed to produce an additional barrel (of oil or gas). Marginal costs rise for a variety of reasons. Most notably, as a field nears its limit of recoverable reserves (a level of reserves chosen in response to commercial decisions as described below) and more output is sought in a given period, additional production will be from more distant areas where natural reservoir pressure will be lower and extra (costly) artificial pressure must be induced.[1] Managerial, operational and logistic diseconomies will also cause marginal costs to rise.[2]

Figure 2.1

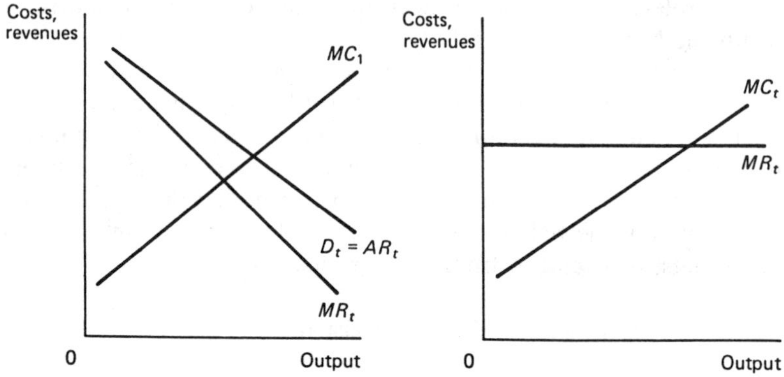

(a) If monopolistic
MC = Marginal cost
MR = Marginal revenue

(b) If competitive
D = Demand
AR = Average revenue

These revenues and costs can be presented as in Figure 2.1. As the range of options for production is bound by total recoverable reserves, these curves are only drawn in this region. The marginal operating profits will be the difference between marginal revenue and marginal cost. Both a monopolistic industrial structure and increasing marginal costs imply the marginal operating profit schedule will slope downwards.

For any discovery there will be a sequence of marginal operating profit lines, one in each chronological period. The recoverable reserve capacity links this sequence by constraining summed production. Considering only two periods for simplicity,[3] and assuming the capacity constraint is binding, a decision to expand output now necessarily entails lower production later.[4] In Figure 2.2, increasing production from $q^a{}_t$ to $q^b{}_t$ involves a reduction in later supplies from $q^\alpha{}_{t+1}$ to $q^\beta{}_{t+1}$. After this shift to $q^b{}_t : q^\beta{}_{t+1}$ if more production was brought forward, profits would increase by $P^b{}_t$ and fall by P^β_{t+1} in the margin. Discounting to reflect the company's preference for earlier profits (a preference reflected in the company's opportunity cost of capital), the marginal net operating profit may be ascertained, P^b_t less P^β_{t+1}. Maximisation of profits will occur when the increase in profits now is no longer greater than the fall in later discounted profits (at $Q^c_t : Q^\gamma_{t+1}$). This is a geometric interpretation of the familiar proposition[5] that marginal net profits will rise at the opportunity cost

Figure 2.2

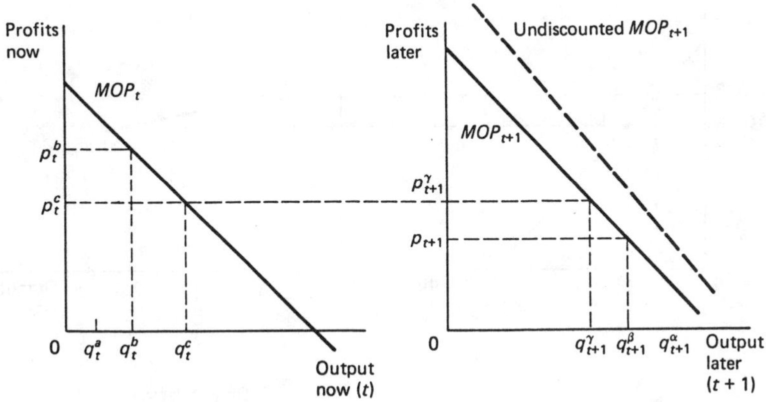

MOP_t = Marginal operating profit in period t

$q_t^a + q_{t+1}^{\alpha} = q_t^b + q_{t+1}^{\beta} = q_t^c + q_{t+1}^{\gamma}$ = Recoverable reserve capacity

of capital. (Note that this reasoning is only valid if the capacity constraint is binding.)

Capacity decisions – the first part of an oil – field decision – will now be considered. The marginal benefits from installing capacity are the marginal operating profits considered above. The company's optimum will occur when discounted marginal operating profits cannot be increased by delaying or bringing forward depletion. The marginal profits from increasing production now would be the same as the marginal profits from increasing production later. Increasing capacity implies production now (and/or later) can be increased without necessarily reducing output in other periods. The optimum marginal operating profits will hence show the marginal benefit from additional capacity.

Depending on the relative slopes of current and later marginal operating profits, an increase in capacity may involve a faster or slower rate of depletion although production in every period will never fall. If marginal operating profits are more sensitive to output now rather than later, the company's optimum profile will move towards a slower rate of depletion as capacity expands (Figure 2.3). The marginal benefit of capacity will be related to these sensitivities. In Figure 2.3, the marginal benefit with capacity *a* will be $OMOP_a$

Figure 2.3

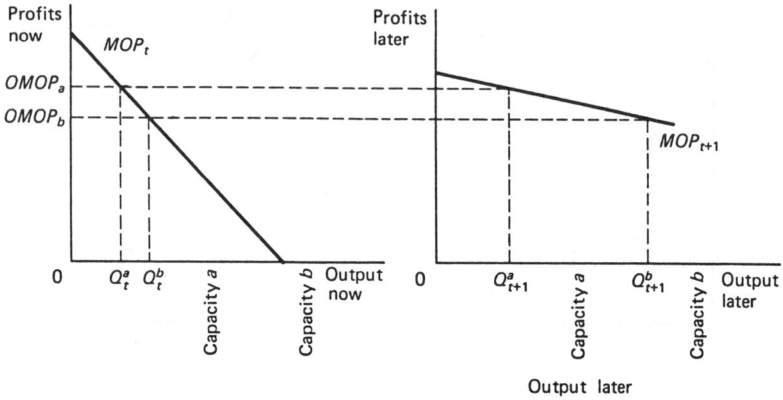

OMOP = Optimum marginal operating profits
capacity $a = Q^a_t + Q^a_{t+1}$; capacity $b = Q^b_t + Q^b_{t+1}$

and with capacity b it will be $OMOP_b$. This can be represented at the time of the capacity decision by discounting the optimum marginal operating profits at the firm's opportunity cost of capital; this is the marginal benefit line in Figure 2.4. The marginal costs of installing additional capacity can also be represented on this diagram. As more and more oil (or gas) is included in the recoverable reserves of a given tract, incremental barrels of recoverable reserves will be further from the centre of highest pressure in the middle of the field, will be harder to extract involving deviated drilling and more separation facilities, and will exhibit diminishing marginal returns. [6] Maximising profits dictates that the capacity or recoverable reserve base should be expanded until the marginal benefit of additional capacity no longer outweighs the marginal costs.

Fitting both aspects of the field decision process together yields a diagram depicting the simultaneity of the development and production decisions (Figure 2.5). Given the solid schedules, the optimum capacity choice is A with a depletion profile of q_t: q_{t+1}. By way of example, consider the effect of greater expected marginal profits in the later period. This would be shown as a higher marginal operating profit schedule such as the dashed line (MOP'_{t+1}). The depletion choice with unchanged capacity would delay production and optimal marginal profits would rise. This would increase the marginal benefits of capacity (to MB') and encourage capacity installation, so long as

Figure 2.4

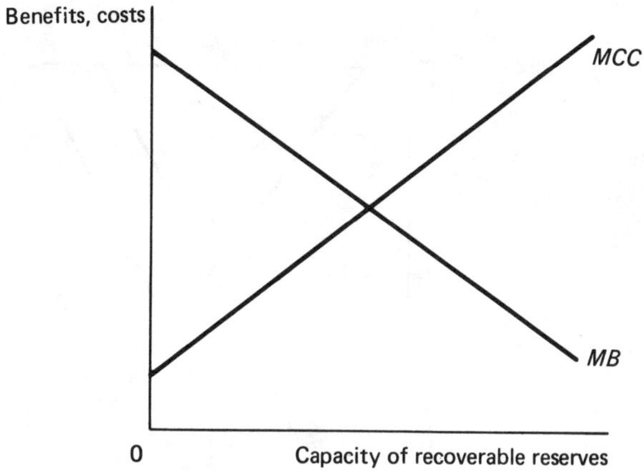

MB = Marginal benefits
MCC = Marginal capacity costs

the marginal cost curve is not vertical. With additional capacity, both output now and output later will tend to rise. Overall later production will increase but the net effect on early production is indeterminate.

A particularly important parameter in development decision is the rate of discount. Increases in the rate of discount will have two effects: first, later depletion will appear relatively less attractive compared with earlier depletion, and second, marginal benefits to installing capacity will decline. Hence, recoverable oil reserves will be smaller and will be depleted more rapidly. Because discount rates used by companies tend to be determined in capital markets, factors which depress or inflate interest rates will impinge on North Sea activity. Clearly though, the rate of interest is a key determinant of many sectors of activity, and to predict the overall effect on North Sea activity the feedbacks through these other sectors need to be assessed. For example, if interest rates are excessive then all forms of investment will be stifled and capital costs, including the costs of North Sea capacity, may decline. The net effect would be far from clear. In any case, the determination of interest rates is outside the scope of this chapter.[7]

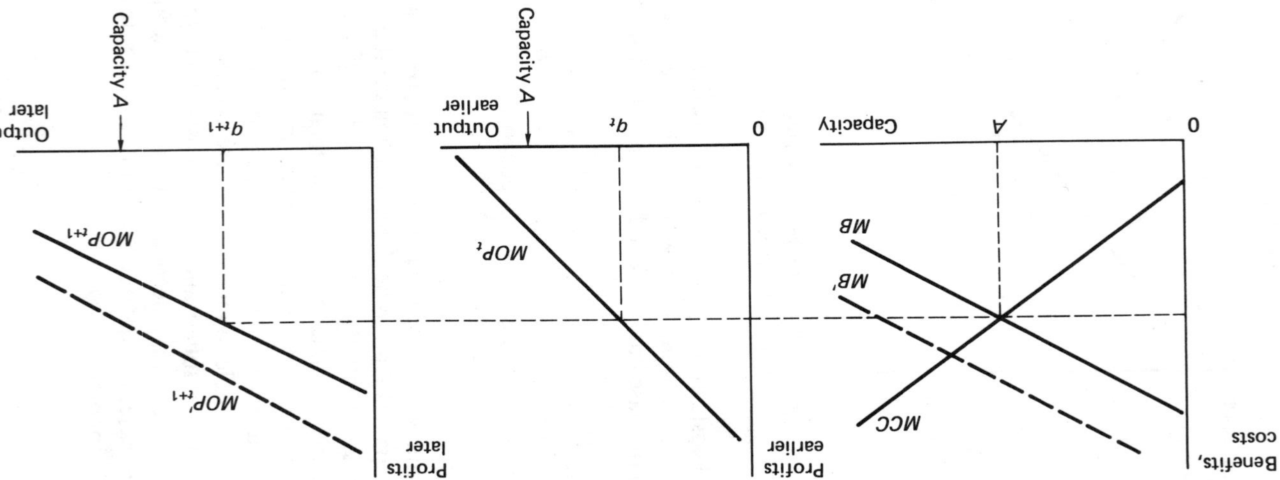

Figure 2.5 Depletion choice: $q_t + q_{t+1} = A$

Note: Marginal operating profits are discounted into the year of decision about capacity size

2.2. THE OIL AND GAS INDUSTRY IN AN UNREGULATED MARKET

Licensing

In an unregulated market we assume the government would establish a profit-maximising body which would sell licences to extract oil in some form of auctioning process. As a result, the technically most efficient companies with lowest costs would tend to outbid the less techinically competent companies and secure licenses. Nevertheless, it may be the case that the pace of licensing would be dominated by pressures for a rapid rate of issue. In any licence round the government licenser would supply licences up to the point where the marginal return from perceived later demand just balanced the marginal return on an incremental licence. There would tend to be a number of factors affecting the pace of licence award in this market.

First, government desires for early revenues may tend to result in later benefits being heavily discounted. Similarly, the licenser's attitudes to risks are relevant. The demand in future periods, although only apparent in the future, affects the marginal return from delaying licence issues and determines the licenser's willingness to offer licenses now. The demand for licenses in the future is uncertain because the factors affecting the industry's bids and the expected values of those factors will be unknown: this is uncertainty which can never be avoided. Additional uncertainty which can be avoided exists, however, since the licenser must guess the industry's views on these factors, and because these views determine the level of the industry's bids. These views, though, relate to information which is known to the industry but can only be estimated by the licenser. The risks associated with deriving expected industry bids are risks generated by the market. A risk-averse licenser would not ignore either type of risk and would seek to avoid them by accelerating licensing compared to a system which does not generate additional risks.

A second feature impinging on the operation of this market is the extent of competition displayed by bidders. If there is ample competition for licences then bidders in the market will tend to find that they will only be able to secure a licence if their bid represents the full value of the rent, otherwise profit–maximising competitors would outbid them. Competition ensures that the sums companies are prepared to pay for licences in different periods are also the sums they are willing to pay.

Competitive forces tend to push bidders to the maximum they are prepared to pay; however, such competition may not be likely in a North Sea context. The scale of costs, the exposure to risk and the heavy front-end loading of costs form substantial barriers to entry for small companies and tend to restrict the offshore industry to an oligopolistic structure. With an oligopolistic market structure, companies will tend to collude and lower their bids below the levels they are prepared to pay. The degree of collusion will be restricted by legal barriers, and by the fear of invoking harsher legal barriers. If the oligopoly and the level of collusive behaviour tend not to change over time, then the reduction in the demand facing the licenser is likely to be similar in different time periods and the pace of licensing will not be affected.

However, the oligopoly is unlikely to persist over time as the barriers to entry which create the conditions necessary for collusion to exist are likely to dwindle over a period of time. In the short term, larger oil companies will tend to face relatively lower costs than smaller non-oil companies. Oil companies possess the experience and have ready access to the expertise necessary to prospect and manage a potential oil-bearing tract. In the long term, though, this expertise will be freer to move to new entrants. Similarly, the reasearch and development information on technological innovations may originate in large oil companies, but it will dissipate to other companies over time. Larger companies may be better suited to bear dry hole risks over a prolonged period, though as an area is drilled and the range of uncertainty narrows, the risk-spreading advantage narrows. Capital market shortages in the initial period may give an advantage to companies with internally generated funds, such as those involved in oil extraction and tending to benefit from the associated economic rent. The fears of shareholders place constraints on small non-oil companies, especially as shareholders tend to believe companies should stick to the areas where the management has some expertise, preferring to diversify their own portfolios rather than purchasing shares in diversified companies. However, once profits become established in a sector the fear may recede. If a small company wants to participate in a joint venture consortium it will face costs of setting up the consortium in negotiations and formalisations; costs which tend to be faced only once and which will be likely to fall over time. When an area is first licensed, established oil companies tend to have sunk fixed costs in overheads and equipment, which again will appear as an advantage over new entrants. Hence uncompetitive behaviour is likely to be more pronouced in the early years of licensing than in the later years.

If the collusion in the early years of the industry is not expected to last, both the industry and the licenser will be affected. The colluding companies may fear that licences not issued now may go to the competitive fringe, since they are no longer guaranteed to receive licences issued later. Hence they will pressure the licenser into a faster rate of issue than might occur if the collusion were permanent or if no collusion existed.

Collusive behaviour on behalf of the licensees may partially offset the tendency of the licenser to issue tracts sooner than later. However, this must be viewed in the context of the scramble for upstream acreage in the 1960s and 1970s which encouraged the licensees to prefer early licensing. Faced with a reduced role in Middle Eastern oil activities, oil companies were particularly keen to be actively involved in other oil-bearing areas such as the North Sea. In a market for North Sea licences, this pressure would have tended to generate higher value bids for current licences, thus bringing forward licence issue.

Exploration

In the absence of policy, exploration activity would be likely to be both deferred and deterred. First, consider why exploration drilling may be deferred. An exploration well generates knowledge concerning the licensee's tract but also provides information about neighbouring tracts and the basin generally. If it is not possible to exclude other companies from the results of an exploration well, the information would be akin to a public good and the market would be reluctant to drill exploration wells. Companies will wait for other companies' drilling results and exploration will be deferred. Conversely, if the information is treated highly confidentially, companies will not release or trade the information and an unregulated market would be likely to generate repetitious exploration activity. Given the possibility of collusion in the early days of the North Sea, on balance it is likely that exploration results would initially be treated as a public good and exploration drilling depressed. However, once new entrants emerged and competition increased, exploration results would become highly confidential and the unregulated market would generate repetitious information.

The factor likely to deter exploration activity is similarly most important in the initial years of an industry when there may be an element of collusion. If companies are required to pay large amounts in licence payment then these sums may crowd out exploration expenditures. In a world of capital market rigidities and a world where North Sea bids might dominate a company's portfolio, the scale of the expected economic rent will not only determine, but will be determined by, licence bids. Successive increments to bids will replace other opportunities of higher and higher value (or at least raise the company's borrowing costs). This, albeit temporary, increase in the firm's opportunity costs will lower the attractiveness of, and the rent expected from, North Sea projects and will deter exploration activity.

Development

Once fields are discovered, recoverable reserves will tend not to be as great at they would be without risks. In the absence of a complete set of forward and state contingent markets, oil and gas producers must base their decisions on expectations of future demand. As demand in a future period will only be realised once the sequence of spot markets reaches the future period, and as state contingent markets are absent, a risk of misestimating demand must be borne by companies developing oil and gas fields.

Even assuming expectations are rational and the central estimate is unbiased, firms will respond to these risks by reducing production in periods of greatest risk and either increasing production in other periods or reducing recoverable reserves. As more distant periods are associated with greater uncertainty, the effect of risks would be to generate a more rapid rate of depletion from a smaller recoverable reserve base. In the absence of policy, markets may develop to spread or pool these risks but, given the scale of these risks, it seems unlikely that the deterrent to production would be eliminated.

Similarly, once fields are discovered, recoverable reserves will tend not to be as great as they would be if North Sea projects were exempt from Corporation Tax (CT). CT would presumably apply to North Sea companies in an unregulated market. CT is a tax on the excess of revenues over historic costs which, since companies are interested in post-tax profits, might affect development and depletion decisions. Indeed, only under rather rare circumstances might CT be a neutral tax. The impact of CT may be identified by considering the CT assessable profits base: if the CT base is identical to company pure profits then the tax will be neutral.

Consider the marginal operating profits which dictate the depletion decision. In the absence of risk, these will be identical to the marginal CT base as viewed by the company. Moreover, since CT would only be payable once profits are positive, it is appropriate to discount them at the company's opportunity cost of capital. With a single tax rate, post-tax profits will be a fixed proportion of pre-tax values. So long as this tax rate is constant over time the proportion will be unchanging. Although post-tax marginal operating profits will be lower than pre-tax profits (implying lower marginal benefits), for a given capacity the post-tax profit maximising output profile will be unaffected.

Marginal costs of increased capacity would be similarly reduced by CT. As capital equipment costs are deductible in computing CT assessable profits, the effective marginal cost of additional capacity would be reduced to the extent that the company is paying or will pay CT. In the case of a company with other profitable activities and with current CT liability, the marginal costs would be reduced as current CT would no longer be payable. With a constant single tax rate and assuming all capital costs can be deducted immediately, the reduction in marginal costs would just match the fall in marginal benefits. Following the reduction of first-year capital allowance in CT not all costs can be deducted immediately and hence the fall in effective marginal costs will not match the fall in marginal benefits. Similarly, all companies would have current CT tax liability. If this field were a company's sole asset, then it would only have CT assessable profits in the later period once the field is onstream. Effective marginal capacity costs would then be reduced by the extent that later CT could be cancelled.

Although all capital expenses are deductive from CT, the delay between capital expense and tax rebate represents an additional cost of the company which is not allowed for in CT legislation. CT only allows for historic costs to be deducted. Effective marginal costs would be reduced post-tax, but not as much as the reduction in marginal benefits. The longer the delay, the greater the real opportunity cost of capital and the faster is the rate of inflation, then the smaller will be the reduction in effective marginal capacity costs. In these circumstances, CT is no longer neutral and will tend to reduce capacity and lower recoverable reserves.

One further consequence in an unregulated market is that the macroeconomic effects of oil developments on the UK economy will not be considered in development decisions.

If there are benefits to be gained from offshore production that accrue to others besides the producers, then less North Sea activity would be likely than if the beneficial externalities were internalised. Alternatively, if offshore production creates problems for the on-shore economy then these problems would create external costs which, if borne by the producers, would reduce North Sea activity. The marginal benefit and the marginal capacity cost schedules would alter if companies respond to these externalities; however, the direction – let alone the magnitude – of the change is not clear.

Intertemporally, external effects may not be uniform. One external cost often discussed[8] is the cost of changing (deindustrialising) the manufacturing economy in response to the balance of payments boost associated with offshore activity. The degree of change, and hence the costs, will tend to increase as the depletion profile is more peaked. If such costs accrued to offshore producers they would opt for a flatter depletion profile than they might otherwise choose.

With an oligopolistic UK gas-producing industry and with international gas trade poorly established, UK firms would find output and pricing decisions related (this is unlike the UK oil industry where oil producers must act as price-takers). If more gas were to be sold in a given period then demand must be stimulated by lowering the price of gas. The marginal revenues of additional gas production would then be below the price of gas – as the lower price would reduce revenues on intra-marginal units of output – and would be negatively related to gas production. Thus, depending on the elasticity of gas demand, the marginal revenue curve will slope downwards: the more inelastic is UK demand then the steeper will be marginal revenue.

Two consequences emerge in a comparison to a gas industry not dominated by oligopolistic forces. First, for any given capacity and for all output levels, marginal revenue and hence marginal operating profits from early production will be lower in an oligopolistic industry. This will influence gas producers to delay production. Second, the marginal operating profits that maximise a firms' operating profits will be lower and hence the marginal benefits to installing capacity will be lower. Less capacity will be installed and recoverable reserves

Table 2.1 Characteristics of an unregulated market

Licensing	(a) too fast
	(b) most efficient firms receive licences
Exploration	(c) too little
	(d) too slow
Development	(e) too little
	(f) too slow
Production	(g) gas too slow
	(h) gas too little
	(i) oil plateau too short
	(j) oil too little
Economic rent	(k) small scope to increase licenser's share from gas fields
	(l) small scope to increase licenser's share from oil fields
Windfall profits/losses	(m) from gas fields accrue to companies exclusively
	(n) from oil fields accrue to companies exclusively

will not be so great. In short, delayed output and lower reserves will result compared to an environment either where competitive forces always exist or where the international gas market dictates UK prices always. When both of these conditions are violated differences emerge.

2.3 DESIRABLE POLICY TARGETS

Employing the outcome of an unregulated market as a benchmark to determine desirable directions for policy, it is apparent that there is considerable scope for improvement by government. Table 2.1 summarises the areas of greatest deficiency in an unregulated market. The two broad areas where policy action is desirable are efficiency and distribution.

Efficiency

An efficient outcome would be characterised by one where the gains from trade dictate economic decisions, and consequently economic activity would be pursued at a level that maximises the value added from that activity.

Concern for efficiency relates to a concern over the total resources used in the oil industry. The extent of North Sea activity should not be so great as to crowd out activities of a higher social value, nor should insufficient funds, time and expertise be devoted to developing oil fields. Similarly, an efficient mix of factor inputs should be used in offshore projects. In the margin, the search for, and the development of, oil fields should involve resource costs that are just matched by the benefits to the society from an additonal exploration well or from an additonal barrel of recoverable oil. A less familiar dimension to efficiency but of great importance to extractive industries, relates to the time dimension of depletion. The concern is that the depletion of a reserve base will occur neither too quickly nor too slowly. One element of the opportunity costs of producing a barrel of oil now is that this barrel cannot be produced at a later date, and so any benefit which might arise later will be forgone. In the margin, the benefit forgone should be matched by the benefit realised now. Otherwise, an increase to total net benefits would be possible by redistributing the time profile of production.

In an environment where the gains from trade alone dictate the pace of licensing, the government would not be able to influence the licenser's behaviour as it might in an unregulated market. A government commitment to resist temptations to direct the licenser would therefore be desirable. Indeed, the government's intention should be contrary to government interests in attempting to reduce the pace of licensing. Similarly, if the government were to (act as if it could) stimulate the development of insurance markets or rectify the imbalance of information between licenser and licensee, the detrimental impact of risks would be ameliorated and the licenser would no longer seek to avoid these risks by increasing the pace of licensing. Furthermore, if the factors of production could have moved easily into other employments, the scramble for access to oil supplies from the North Sea would not have occurred. In this way the pace of licensing would have been dictated more by concern to maximise the value added offshore. However, to the extent that other factors would tend to cause the pace of licensing to be too rapid in an unregulated market, the resouce allocation appears inefficient. An intention to reduce the pace of licensing should then be welcomed as an intention to improve the resource allocation. If exploration activity were determined solely by the desire to maximise value added in the offshore industry, the effect of licence payments in the unregulated market increasing the interest rate and hence crowding out exploration spending would not be relevant. Indeed one may have hoped that either the asset would be identifiable and would provide sufficient collateral for the lender not to feel exposed to default or that joint venture bids and joint venture loans would emerge to scale down the impact on the debt market.

Similarly, the public good attributes of exploration drilling suggest more drilling would be necessary to achieve the maximum value added. Hence exploration in the unregulated market will tend to generate too little drilling and if additonal resources were transferred out of their marginal use and into North Sea exploration, the net value of the output would not be as great as the net value gained.

A government intention to encourage exploration, particularly in the initial period when the public good nature of drilling will be strongest, would be welcomed.

Private sector development and production decisions in an unregulated market are likely to result in less North Sea capacity and lower recoverable reserves that would be necessary to maximise the value added in this activity. If specialist insurance companies were to emerge to shoulder financial risks, if externalities in marginal costs or marginal benefits were to be internalised and if competition forces in the gas industry were to regulate any oligopolistic pricing tendencies, additional investment would be forthcoming.

An intention to encourage the development of North Sea resources may be seen as a way of achieving the additional investment while bypassing the market-related transaction costs which allow these factors to appear. In this way resources would be encouraged away from low value added uses and towards higher value added uses, thereby improving efficiency.

Depletion profiles for North Sea oil in an unregulated market are unlikely to appear efficient. The direction of imperfection in gas supplies is likely to be different from that in oil supplies. The initial isolation of the gas industry from international competition will tend to delay gas supplies compared to the depletion profile that maximises value added in gas production. In the oil industry, aversion to risks which affect individual companies but which can be spread or pooled across the industry and failure to consider the macroeconomic externalities both tend to encourage more rapid and more highly peaked oil depletion than would occur if the profile were efficient.

To overcome these inefficiencies in the allocation of resources over time, an intention to speed up gas depletion and to slow down oil depletion would be justified. Given the undersupply of both gas and oil investment, this implies early gas production and later oil production should be encouraged.

Distribution

Apart from reasons relating to efficiency, policy intention might also emerge for reasons of distribution. In an unregulated market, the economic rent from offshore activity would be distributed between the licenser and licensees according to the extent of competition in the market for licences; that is, with less competition more rent would be captured by the licensee, and less would accrue to the licenser. A wider perspective will show how these benefits or losses might be passed on to oil and gas consumers, to general taxpayers or to shareholders of oil and gas companies. Consider first the licenser's rent disposal. Presumably X-inefficiency and organisational slack would emerge within the licensing authority, but probably not on the same scale as the scale of rent generated and only insignificant sums would be absorbed by the licensing entity. The remainder would be passed on to the politico-bureaucratic establishment, presumably in an identifiable form with distribution under government supervision. While a portion of the rent is likely to be absorbed in the political and bureaucratic machinery, so long as it is kept as a separate and identified sum then the majority of the rent is likely to pass through the machinery. The beneficiary will then tend to be the general taxpayer. Next consider the licensee's rent disposal. As competitive forces replace oligopolistic practices then the oil companies will face reduced pressures to pass profits on in dividend distributions and the disposal of rent by private sector companies will tend to be more beneficial to shareholders (especially in the short term) and less beneficial to employees. In summary, if oligopolistic forces dominate the licenser/licensee relationship then oil company employees are likely to benefit most whereas, if the industry is competitive, the licenser and hence the general taxpayer are likely to receive the greatest benefits.

So far we have dealt with the distribution of economic rent. The distribution of any windfall profits or losses should also be mentioned. In the unregulated market, the licensing system would both protect the licenser from windfall losses and exclude the licenser from any windfall gains. With licences paid for in cash bonuses, no further payments arise in later years.[9] CT, however, will tend to reflect windfall changes and though windfall profits or losses will not be felt by the licenser, Treasury revenue will change with any windfall.

On equity grounds, a more evenly dispersed distribution of income and wealth may be supported.[10] Application of this principle to the

North Sea would favour wider groups receiving economic rent. Moreover, a system which distributed windfall profits to the less wealthy while leaving windfall losses with those of greater wealth, would be a system supported by this equity principle.

2.4 GOVERNMENT INTENTIONS

Table 2.2 provides a brief outline of the development of North Sea oil policy. Government intentions and policy targets associated with various policy decisions are listed in the left-hand column of Table 2.2. An initial government intention was to use North Sea oil to provide a fillip for depressed Scottish and North-east English industry by encouraging rapid and thorough exploration offshore. Increasing the involvement of UK firms in supplying offshore capital equipment and the percentage of UK ownership of licensed acreage were shorthly added to the list of government intentions. Statements supporting an intense period of development to secure early self-sufficiency and a rapid build-up in supplies emerged once it became commonly accepted that the reserves in the UK sector of the North Sea were not insignificant. Following the discovery of commercial deposits of gas, the government stated its intention to collect the gas profits from the oil companies (and implicitly to pass these on to consumers of gas). After the rapid increase in crude oil prices during the Yom Kippur War had enhanced the commercial viability of the more northerly oil deposits, political opinion moved in favour of collecting the large oil profits. Although history is not clear, it seems likely that there was political agreement over the intention to capture the profits despite differing party intentions over where to distribute the gains. In Conservative eyes the intention seems to have been to benefit all income taxpayers by lowering the level of income tax, while in Labour eyes it was apparently to benefit users of public services by raising public spending. More recently, as oil production has expanded to self-sufficiency, government attitudes have tended to support a flatter production profile and a longer plateau of oil supplies than might otherwise occur.[11]

Interpretation of Policy Intentions

The second column of Table 2.2 records our economic interpretation of each of the government intentions by associating government

Table 2.2 Government North Sea policy intentions and instruments

Government targets	*Economic interpretation (letters from Table 1.1)*	*Instruments of policy*	*Impact of Policy* — *Objectives aided*	*Objectives hindered* (*letters from Table 2.1*)
Linkages to British industry	(e)	Discretionary licensing (British ordering)	short-run (e)	long-run (e)
Rapid and thorough exploration	(c), (d)	Discretionary licensing (work programmes)	(c), (d)	(k), (l)
UK (and State) ownership	(k), (l)	Discretionary licensing (especially via the British National Oil Corporation)	(l)	(i), (b)
Rapid development to self-sufficiency	(f), (g)	Discretionary licensing (favouring companies developing UKCS)	(f)	(e), (a)
		Approval of development plans		(j), (h)
Collect gas profits	(k)	British Gas Corporation	(k)	(k)
		Monopsony	collection from companies	transfer to consumers (g), (e), (h)
Collect oil profits	(n), (l)	Oil taxation (BNOC)	(n)	(e), (j), (i)
Lengthen plateau production	(i)	Staged approval of development plans, depletion legislation and guidelines	(f)	(i), (e), (j)

targets with the economic factors shown in Table 2.1. Here we do not seek to establish that the fulfilment of an intention will improve efficiency, but merely to ascertain the general direction policy must necessarily take. Nor do we consider here whether the existence of a justifiable intention is by design or by coincidence: this is a topic of the economics of politics. We merely wish to assign an economic interpretation to each of the government intentions noted above.

Although UK ownership might be justified for efficiency reasons – since this would tend to support the Scottish supplying industry – it can more accurately be described as an intention to affect the distribution of the economic rent associated with offshore fields. Rapid exploration and developement of self-sufficiency, subsequently with a flat production plateau, are presumably intended to resolve time inefficiencies. The intentions to collect gas profits and to collect oil profits seem to have arisen to fulfil different objectives. With gas the concern seems to have been to collect economic rent, while the intention for oil is apparently to extract windfall profits from existing developments as well as collecting economic rent on new fields. Thus the distribution may be enchanced if the profits leave one segment of the international economy (the oil multi-nationals) for UK groups.

There are two additional targets not included in the list of government intentions above. First, no explicit recognition of the inefficiencies associated with the rapid pace of licence issue in the unregulated market is visible in government's policy intentions. The source of pressure for too rapid a pace of licence issue in an unregulated market, however, emanates from the government's desire to speed up the collection of licensing proceeds. One consequence of choosing a discretionary licensing system is to eliminate such pressure and hence, having chosen not to auction licences, a government intention to reduce the pace of licensing is no longer necessary. Second, an additional target to encourage the production from oil and gas fields and an intention to promote more offshore development activity would be welcome. This encouragement and promotion would increase the recoverable reserve base but would not be directed at the timing of developments or the rate of depletion. We believe this to be a significant omission and to be indicative of the government's lack of appreciation of the need to encourage offshore developments.

2.5 POLICY CONSEQUENCES

The government targets above do not seem to be misplaced. Targets however, are only one aspect of policy and there are many significant constraints acting upon government which inhibit the achievement of policy targets.

Indeed, our examination of the impact of the instruments of policy suggest that, even if a justifiable direction for policy has been ascertained, it is a complex task to move resources in the direction.

Policy Instruments

In the third column of Table 2.2, the policy instruments which seem to have been chosen to fulfil each target have been listed alongside the targets. One instrument, discretionary licensing, has been used to fulfil many objectives. The different dimensions of discretionary licensing are separable and can be assigned to specific targets. Thus the preference in discretionary licensing given to companies ordering British-built equipment was presumably designed to achieve the target of boosting Scottish industry and the North-east English economy, while the decision to assess companies by work programme (instead of cash)[12] apparently ensured rapid and thorough exploration. The target of UK state ownership was an objective of the licensing procedure also, in particular the discretionary preference for groups of licensees with UK involvement.[13] One further, albeit minor, function of discretionary licensing was the hope that development to self-sufficiency might proceed rapidly if there were no licence payments. The build-up to self-sufficiency was also the target of the Department of Energy's depletion control mechanism which let company development plans proceed virtually unhindered. More recently, once several oil fields had been developed, regulations were set up to moderate the pace of depletion. The legal mechanism for adjusting planned output was instituted in the Petroleum and Submarine Pipelines Act 1975, which has since been clarified by the Varley Guidelines of 1974, a statement by David Howell made in July 1980 and the so-called 'Lawson reassurances' of 1982.[14] Planning approval can also be used as an instrument of depletion, especially since consent is now granted in stages for a limited period of production only. This network of control has yet to be implemented as a method to lengthen the plateau period of production by reducing the plateau rate output. No concern over the rate of gas depletion (apart from the flaring of associated gas) has been evident.

Collection of the profits from gas production has been the target of one instrument while different instruments have been used to collect the profits from oil production. Setting up the British Gas Corporation (BGC), formerly the Gas Council, as the monopsony for gas was designed to transfer the profits emanating in gas fields away from the oil companies. Profits from oil production have been the target of special tax regulations, regulations which have changed many times since 1975 in attempts to collect an increasing share of an increasing amount. Another instrument to collect oil profits was the British National Oil Corporation (BNOC) which enjoyed, until the 1982 privatisation of BNOC's production and exploration activities into Britoil, a share of oil profits through its equity stake in oil fields. This equity stake was built up via privileges granted to BNOC in the fifth and sixth licensing rounds. In the fifth round BNOC was given a 51 per cent share in all 44 blocks allocated. In the sixth licensing round oil companies had to compete for tracts by giving BNOC equity interests of more than 51 per cent, by giving BNOC options to purchase or sell oil and by 'carrying' BNOC's interest.[15] By privatising Britoil the government claimed 51 per cent of the expectations of the present value of BNOC's exploration and production interests.

Policy Impacts

We now wish to analyse the impact of this set of policy instruments. By comparing this impact to the unregulated market, we may assess the performance of policy. We have already suggested that policy intentions are roughly appropriate; if however, policies do not fulfil the roles they have been assigned, then the policy instrument may not be appropriate. Further, if there is an inherent weakness in the design of policy such that appropriate instruments will never be forthcoming – perhaps because of political or bureaucratic motivations – then no matter how suitable the policy intentions are, policy should not be applied. Indeed the assessment of policy instruments tends to be relevant than the assessment of policy intentions.

The format of the next sections will be to explore how each instrument has affected its target parameter and how each instrument may have had unintended consequences on other variables. Policy is considered under three broad headings: licensing, depletion control and collection of economic rent. The final two columns of Table 2.2 indicate which objectives from Table 2.1 are aided and hindered.

Licensing

Discretionary Licensing and Work Programmes

Since the criterion in the licensing system that is (and was) designed to encourage rapid and thorough exploration is the work programme criterion, the impact of this aspect of policy may be assessed by its effect on the pace and extent of exploration drilling. If more exploration activity would be likely than in the unregulated market, allocating licences in accordance with companies' proposed work programmes will come closer to achieving a faster rate of exploration than if licences are allocated by an auction system.

In making this comparison between policy and the unregulated market we must be aware not only of the impact on exploration but also of the repercussions on other variables. First, though, consider the extent of exploration for any given licence. Clearly, for any given licence, the policy is unlikely to reduce the number of planned exploration wells. In an auction system, companies will plan a work programme which would tend to be the minimum work programme proposed in discretionary licensing. If a company anticipated rent from this licence, then in a competitive auction system the company would raise its cash bid. In a discretionary system where greater work programmed tend to secure licences, a company anticipating rent would tend to expand on its minimum work programme if this would secure the licence. Rent would be sacrificed for additional wells, thus achieving the policy objective. A company, however, will only need to sacrifice rent if it is in competition with other companies.

Given possible collusion in the North Sea industry,[16] proposed work programmes will tend to be close to the minimum programme. Indeed, far from alleviating the problems of collusion, bidding in terms of exploration wells may create conditions such that North Sea companies will find it easier to inhibit new entrants and to maintain their oligopoly. Only established North Sea companies will be sufficiently sophisticated to formulate bids in units of wells. What competition there is in the North Sea industry may be suppressed and (*ceteris paribus*) oil companies will be likely to retain a higher share of the rent. Exploration drilling, though, will not be reduced for any given licence.

Although incentives for companies to increase their exploration drilling may exist, there is little to ensure the additional wells on a given licence will occur earlier rather than later. The sole factor that may encourage earlier exploration on a particular tract stems from discretionary preference in licensing. However, since in the absence of competition the number of additional wells is unlikely to be great, the inefficiency in the allocation of resources over time is unlikely to be markedly improved.

Turning to the second stage of analysis, we find it is less easy to ascertain the impact of discretionary licensing on the variables it was not intended to affect, such as the pace of licensing. If the sole criterion for licensing was the number of exploration wells and competitive forces could be relied on to force companies to bid (in exploration wells) the full value of rent, then a slower rate of licence issue would be likely than in the unregulated market. Maximising the total number of wells will occur where the marginal number of wells from an additional licence issued now is just matched by the *undiscounted* marginal number of wells from delaying the licence issue. Hence later issues will no longer be discounted and the licencer will tend to delay issuing licences. Although the licencer will probably not wish to maximise the total number of wells without any preference for earlier rather than later drillings, the preference will be dictated by the bureaucratic process administering the licence issue. Since the time until the next election is a maximum of just five years, the bureaucracy is likely to have a longer time perspective than government and hence a lower rate of discount. Thus, by distancing themselves from the licensing process, politicians remove one of the pressures creating an inefficiency in the unregulated market. The lack of competitive forces may offset this reduction in the licensing pace, but it is still likely that the too-rapid rate of licensing possible in the unregulated market will, at least in part, be alleviated by discretionary licensing. It is worth noting that this alleviation is despite the implied political quest for speed and seems to have been largely unintentional.

One other impact of discretionary licensing by work programme is on the distribution of economic rent. Once more we find that this will be crucially affected by the level of competition, for if competitive forces are strong then all the rent would be bid in additional exploration wells in each company's planned work programme, and all the rent would leave the private sector companies. We have, however, already indicated that competitive forces will not be aided by discretionary licensing and that the conditions for increasing collusion will abound. A greater share of economic rent will thus be held by the licensees than in the unregulated market. In short, because discretionary licensing does nothing to alleviate the problem of collusion, it cannot be expected to produce an efficient outcome while collecting economic rent. Instead the additional exploration wells on any given licence and the slower pace of licensing will have been achieved at the expense of collecting economic rent.

British Ordering Preference in Discretionary Licensing

Oil companies have a tradition of ordering capital equipment from their home country, which in the case of the North Sea would have favoured US suppliers. In order to break into this market, British suppliers have been favoured by the preference in discretionary licensing given to oil companies ordering British equipment.[17] By raising the demand for Scottish and North-east English construction facilities, some unemployment should be soaked up and otherwise idle resources will be utilised. Neither the extent of additional utilisation of British resources – rather than switching from one employment into another – nor the importance attached to this criterion relative to the other discretionary licensing criteria are clearly definable, but both should be positive.

It may be the case that too little North Sea development expenditure would be forthcoming in an unregulated market. One reason would be that the social costs of construction facilities are less than private costs. Switching from non-UK to British suppliers would be a mechanism to achieve more North Sea development activity. Other features causing too little development expenditure remain. In addition, one may query if the boost to British industry will last any longer than the duration of the policy. If British suppliers become X-inefficient as a result of receiving an advantage over their competitors, then once the advantage disappears they will find it hard to compete against international suppliers. Indeed, in the long run it may be best to ignore short-run devices to support industries whose long-run future lies in an international market. More fundamentally, if there are problems in the British economy which create involuntary unemployment, then these problems should be tackled directly. Indirect attempts to manipulate demand will tend to be fruitless if the demand–supply equilibrating mechanism does not operate.

Finally, if oil companies are forced to order British when they would prefer not to, this suggests that British firms will not be the minimum-cost suppliers. The cost of development North Sea fields will tend to reduce the number of offshore developments and represent a move away from the desired outcome to the detriment of economic efficiency.

Disçretionary Licensing and State Ownership

By offering to include BNOC (and the other State organisations operating in the North Sea) as a participant, groups of companies have received preferential treatment in the discretionary allocations over the treatment received by companies who make no such offers. Thus, if there is rent to be earned, a company may have to sacrifice some of its expected profits in order to secure the licence from the closest other applicant. Since competition between applicants is unlikely to be marked, the sacrifice is unlikely to be great. In principle, a company would probably be fairly indifferent to the choice of its partners, being concerned more with a possible partner's willingness and ability to take on new projects than with its ownership. Partnering BNOC (or BGC) would not seem to be particularly onerous. This does not, however, suggest that economic efficiency will be unaffected. In practice, the pressures encouraging the very rapid expansion of BNOC seemed to come to the point where its ability to take on new projects was affected. Some overall delay in developments has tended to be an unintended consequence. Unfortunately, encouraging BNOC by giving it certain advantages is unlikely to enhance the rather weak competitive forces in North Sea licensing. Perhaps the monopsony nature of the industry's interaction with its suppliers might be weakened, but since BNOC would join groups of companies, the number of groups ordering equipment would not be increased.

More worryingly, it is not clear that an organisation controlled and owned by the State would be efficient at North Sea operations. On the one hand, the risks, costs and lead times inherent in North Sea projects are peculiarities of oil operations that oil (not State) companies have developed an expertise in handling. On the other hand, BNOC may be used – or at least may fear it will be used – to implement North Sea policy. Acting as either an advisory or a regulatory body can only detract from its operational efficiency.

The use of the State oil company as an instrument of policy is no longer possible since the privatisation of 51 per cent of BNOC's assets (in Britoil). Perhaps, though, one factor undermining BNOC's profits has been replaced by another. Britoil is under considerable pressure to prove itself as an oil company, and is keen to diversify internationally and downstream to such a degree that projects which other oil companies might not pursue may be taken on.

Discretionary Licensing, the Favouring of Companies Developing the UKCS and Procedural Approval of Development Plans

In order to encourage rapid development of gas and oil fields, companies developing projects received favourable consideration in subsequent licence rounds. Together with this incentive, the Department of Energy raised no substantial objections to company development plans. If companies anticipated rent on *subsequent* licences then the former instrument might persuade them to sacrifice profits now (by developing faster than they might otherwise choose) in order to secure later licences. In practice, though, it is doubtful if either instrument had the intended effect.

Without competition a company would not be pressured into altering its plans by threats of not being favoured in subsequent licence rounds. Similarly, granting rapid procedural approval of development plans is unlikely to speed up the developemnt process compared to an unregulated market unless a reversal of the procedure is expected later on. If the reversal is expected, then fields may be brought forward. However, unless the reversal is imminent, companies will not find it worthwhile to alter the development timing. Since we are considering discovered fields the choice will only be for one or two years forwards or backwards. Instead, the demand for licences will shift forwards, thus hindering an already inefficient licence procedure. Moreover, additional uncertainty will have been created so that fewer fields will be pursued overall.

Summary on Discretionary Licensing and the Technically Most Competent Companies

We have commented that discretionary licensing may not be able to affect North Sea operations if competition is not rife. Perhaps this is fortunate because the impact would probably be at the expense of efficiency. This is because there is no guarantee that the companies who are efficient explorers, who are willing to buy British, who agree to participate with State-owned organisations or who wish to develop North Sea assets rapidly, would be the most competent firms suited to developing North Sea fields.

Oil Depletion Controls

Production cutbacks and development delays are the two instruments that can be used by government to reduce the rate of depletion. Most obviously, if production cutbacks were imposed, inefficiencies would be likely as capacity would stand idle which, in short term, oil companies could do little to offset.

Development delays, though, encounter immense forecasting problems and are rather inflexible because of long lead times. Below, we focus on production cutbacks; however, our later comments are relevant to both types of instrument.

Impact of Depletion Controls on Depletion Profile: Costs Committed

Implementing controls on production (rather than development delays) will restrict current production. Given the marginal operating profits (marginal revenue less marginal cost) in later periods are still positive for the remaining capacity, then the reduction of output in the earlier period will be taken up by an increase in output in the late period. Only if marginal operating profits are not positive for all the remaining capacity in the later periods will the reduction of output not be fully taken up. This, however, is unlikely.

Impact of Depletion Controls on Depletion Profile: At the Planning Stage

Even in the absence of strategic behaviour (below), companies will tend to alter their planned production profiles if the date of regulation is sufficiently far in advance for their actions to be effective. This, for example, may arise if the government attempts to reduce the peak of North Sea supplies in order to reduce the post-peak rate of decline. Consider the three-period output profile in Figure 2.6. In the absence of regulations, profit maximising decisions would result in the output profile $q_1 : q_2 : q_3$. If control is set for the second period to a level q_2^c then the company will wish to transfer the difference between q_2^c and q_2 into both earlier and later periods. With the control, maximisation of profits will occur where the marginal profits

Figure 2.6

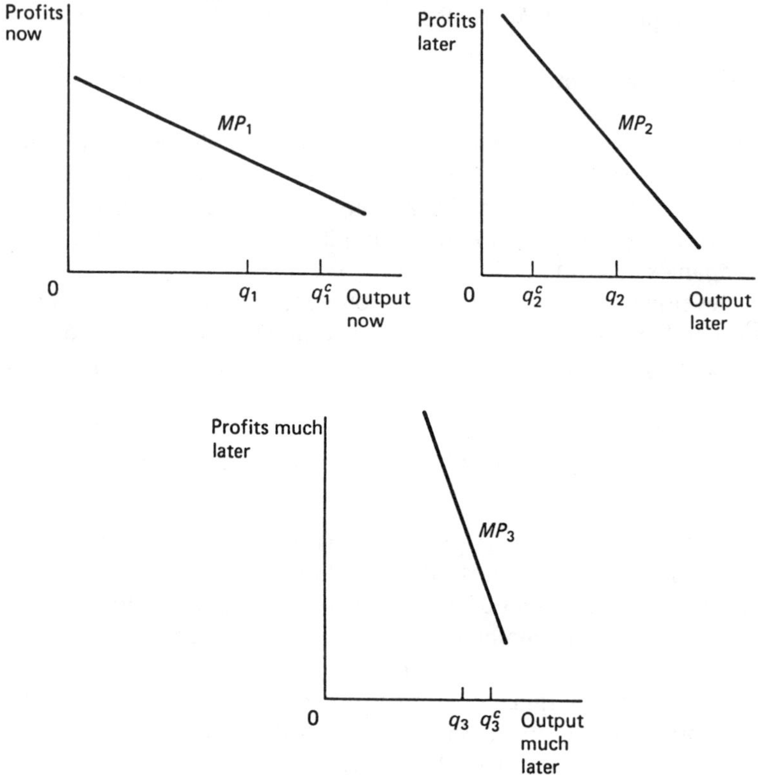

MP = (Discounted) marginal profits

in periods one and three are equated, and production in both these periods is likely to rise. If the marginal profit line in the first period is flat relative to the marginal profit line in the third period, then the bulk of the controlled output from the second period will be brought forward. Since reservoir pressure tends to be higher earlier on, the marginal profit line is likely to become steeper over time so that the fear of depletion controls will tend to encourage the rapid development to self-sufficiency, thereby achieving one of the desirable policy intentions in Table 2.1.

A less desirable consequence, however, seems more likely. Given that the fear of controls will be highest when measures are enforced, the first of the three periods will tend to coincide with the peak of

Figure 2.7

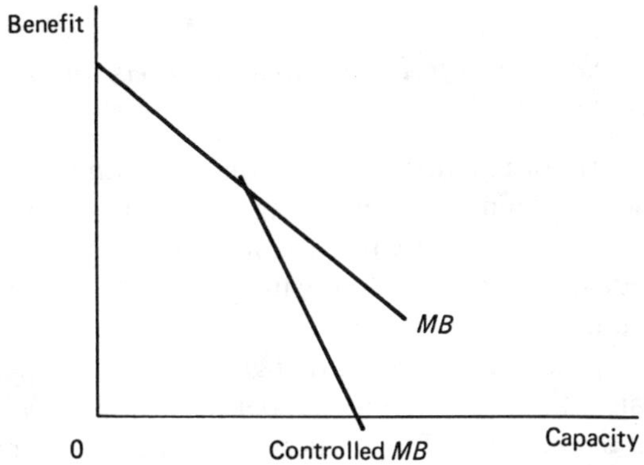

aggregate North Sea production. That is, companies bringing new fields onstream when total oil supplies are peaking will observe (the threat or the actuality of) depletion controls on other fields and would adjust their depletion plans accordingly. In aggregate this will tend to raise the peak of oil supplies and aggravate the time inefficiency which the measures were designed to cure. This phenomenon may well characterise contemporary oil depletion decisions in the private sector.

Impact of Depletion Control on Capacity Levels for New Fields: Non-Strategic Reactions

The impact of depletion controls on marginal benefit schedules indicate the effect on capacity decisions for new oil fields where capital costs are not sunk. If output in the early period is to be regulated within some limit, once this level of output has been reached additional profits will only be available if production in later periods is increased. Because there is no longer an option to increase early production, marginal costs will rise faster and a linked marginal benefit schedule will result (Figure 2.7). Alternatively, output in the early period may be regulated as a fixed proportion of planned output in that period. In either case, operators who expected new fields to be regulated will tend to reduce the amount of installed capacity.

Similarly, because depletion controls affect an oil field's financial prospects and because depletion controls are determined in a political context, an added degree of uncertainty will be attached to oil field investments. This additional uncertainty *per se* will also reduce installed capacity.

Impact of Depletion Controls on Capacity Levels for New Fields: Strategic Reaction

The second form of reaction to depletion control may be in terms of altering a field's planned output. If an operator is certain that production in the early period will be restricted to a set proportion of planned production then the plans can be upgraded to accommodate the regulation. In an extreme case, for a 20 per cent cut (the maximum proposed under the Varley guidelines) the operator might submit a plan with production at 125 per cent of the level that would be chosen in the absence of regulations. When the cut is enforced the operator would be able to produce 80 per cent of 125 per cent.

Such a strategy, however, could not be pursued costlessly by the oil company. In order to give the inflated production plans an air of credibility, additional capacity would have to be installed. The solid line in Figure 2.8 shows the marginal benefits in the absence of regulations, which will also be the company's desired marginal benefits when the plans are controlled. For this latter result, the company must act as if its marginal benefits would be higher in the absence of controls (the dashed line) and hence instal a level of capacity greater than *A*. If the regulators believe the marginal capacity costs follow the dotted line then additional costs to the company will be small and this strategy may be pursued. If the regulators are technically competent, then the additional capacity necessary to substantiate the inflated plans may be too costly for the stategy to be worthwhile. Moreover, the companies would never be certain that controls would be implemented. Such strategic planning would involve a risk that controls might not be imposed. If the company is bound to follow its plan it will then deplete faster than it would like to do.

Impact of Legislation for Depletion Control

Depletion controls will thus tend to affect North Sea activity. Recoverable reserves will be reduced because of the decline in commercial attractiveness while strategic behaviour might result in the installation of excessive capacity to produce a lower volume of reserves at a more rapid rate. In particular it is the belief and the expectations that controls will be implemented which drives companies to change commercially based decisions. Hence setting up the apparatus of

Figure 2.8

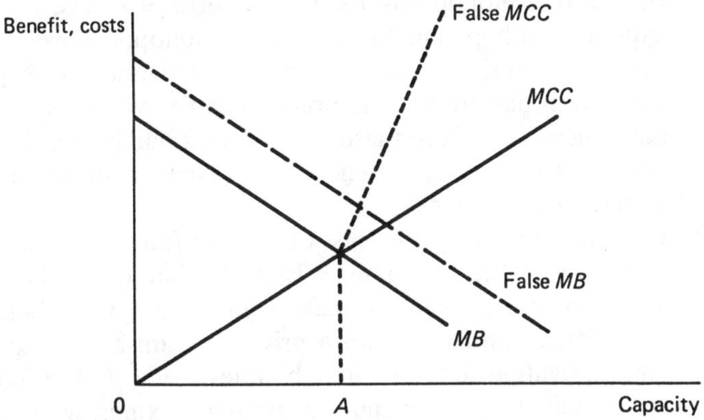

control is likely to have the above impacts even if the measures are never invoked. Once the measures are implemented then a further consequence might emerge.

Impact of Invoking Depletion Controls

When the depletion control apparatus is used to set production levels then the output decisions become the responsibility of administrators, at least temporarily. Once under the responsibility of administrators, if prices or costs change unexpectedly then the signals informing decision-makers that the production profile would need amending would not be forthcoming. This lack of marginal encouragement might create further inefficiencies and accentuate the possibility of changing the depletion profile away from a desirable target.

For example, if the intention is to flatten a predicted hump in production, then field-by-field output levels may be administratively reduced. If expectations change towards faster growth in oil prices, commercial decisions would tend to defer output and automatically smooth the hump in supplies. With output levels set by bureaucrats, there is no mechanism to transfer the need for output adjustment into any change in the level of regulated output. This criticism does not apply to depletion control in principle – since depletion control via price manipulation with taxes or subsidies would not block the signals – but it does apply to the rigid mechanism considered as the instrument of depletion control in the UK.

Collecting Gas Profits

By creating BGC's monopsony rights for North Sea Gas, it seems that the government hoped first to negate the monopoly power of the oil industry to extract the consumer surplus, and second, to recapture the economic rent passed on to the oil companies via the system of descretionary licensing. Once these items of rent had been collected, the intention was apparently to benefit both taxpayers in general and gas consumers in particular.[18]

Before commenting on the success of BGC in fulfilling these aims, it is necessary to establish the impact BGC has had on field-by-field decisions and for this we need to make some assumption about the behaviour of BGC. Since it is not a private organisation, a profit–maximising motivation for BGC may be questioned. Other motivations such as budget, sales revenue, or output maximising are often suggested for State corporations. Profit maximisation, however, is still likely to characterise BGC action in trading North Sea gas: not for profits *per se* but for the organisational slack and X-innefficiency that profits can accommodate. By acting as if it is maximising profits, the BGC can maximise organisational slack.

BGC's Impact on Output Profiles

Accepting that BGC will act as if it is maximising profits, the impact of BGC on a monopsony depends largely on the price elasticity of demand both now and in later periods. Given a decision on capacity, the choice of output profile by a commercial company will be set once the prices have been dictated by the BGC. This commercial response will constrain BGC's freedom to manipulate prices, but otherwise its monopsony rights give it a most powerful negotiating position.

BGC's discount rate is likely to be lower than the private sector's rate, reflecting its narrower alternative range of opportunities and hence lower opportunity cost of capital. With a lower discount rate gas supplies will probably be delayed, as later net operating profits will appear relatively more attractive to the BGC than to private sector companies in an unregulated market.

Another factor tending to delay gas supplies has been the ability of the BGC to prolong the temporary monopoly in gas sales to a more permanent monopoly. Because of its control over the gas distribution network, the BGC has chosen to restrict North Sea gas to the British market. Rather than facing the flat marginal revenue curve implied by increased competition that private companies would face in an unregulated market, the BGC would be aware that the size of its production in the later period would affect the price in that period. Since the later period will be when supplies are declining, demand would exceed indigenous output and some gas would need to be imported. This implies that at the price that would prevail in the absence of BGC being the sole buyer of UKCS gas, the demand for gas would exceed North Sea supplies. At this point on the demand curve, marginal revenue to the BGC will tend to be greater than the price in the unregulated market and this will act as an incentive to delay supplies.

Our conclusion may be surprising. Despite the very rapid expansion in gas that occurred in the late 1960s and early 1970s,[19] we argue that this expansion was not as rapid as it would have been in the absence of BGC's monopsony nor as rapid as would have been desirable. Given the cheapness of gas relative to demand, the conversion to utilise new discoveries could and should have proceeded more rapidly than it did.

BGC's Impact on Capacity Decisions

The argument of the previous section about prolonging the monopoly in gas sales has the effect of lowering the marginal benefits from installing capacity compared to the marginal benefits described in the unregulated market. Since the marginal capacity costs will reflect average costs to the BGC, BGC's marginal capacity buying costs will be above the marginal cost of a competitive company. Both lower marginal benefits and higher marginal costs imply less capacity will be installed.

Perhaps more significantly, the price offered by the BGC for a return on capital will only be relevant for capital equipment yet to be purchased and installed, that is, for avoidable capacity. Unavoidable capacity – reflecting equipment installed and other sunk costs – need receive no inducement from the BGC and no price need be offered for a return on these investments. In particular, BGC's monopsony was enforced after several fields had been discovered between 1965 and 1967. Thus the risks and costs of exploration and initial appraisal drilling were borne by the company without being represented in avoidable costs. For the fields where costs are sunk, the depressing effect of BGC's monopoly on new capacity investment will be offset.

For newer fields, however, companies will be reluctant to undertake exploration and initial appraisal when these costs will not be relevant in the process of price discrimination.

The Long-Run Impact of the BGC on Capacity Decisions

Across fields, the BGC does not have to offer a similar price for gas supplies and can instead act as a price discriminating monopsonist. Since the BGC will effectively be able to offer points along the industry's marginal capacity cost schedule, its buying costs will be identical to the industry's marginal costs, and the distortion to company decisions from BGC's monopsony *per se* will disappear in the long run. As we have noted, however, the BGC's marginal costs will be the marginal avoidable costs. Since the private sector must discover a field before entering price negotiations with the BGC, the negotiated price will not cover its exploration expenses. Although the BGC must have been aware that exploration would dwindle and hence its later profits would decline, it has continued to adopt a field-by-field approach, denying any connection between the price paid on one field with prices elsewhere.[20]

Our analysis suggests that a field-by-field approach might have been a rational profit-maximising policy if the explored acreage was large relative to unexplored acreages. Then the advantage of extracting profits from sunk costs would outweigh the loss of profits from regions that might remain unexplored. This, however, is not an accurate reflection of the North Sea Southern Basin where substantial prospective acreage remains to be explored.

BGC's Impact on the Distribution of Economic Rent

The economic rent from gas fields is split between the BGC, the private sector and gas consumers. Depending on the final destination of BGC's surplus, the consumer surplus may be greater or less than in the unregulated market. If the BGC chose to subsidise gas consumers by lowering the price of delivered gas, the consumer surplus would be greater. The BGC, however, would then need to adopt some non-price mechanism to ration supplies. Since such a non-price rationing mechanism has not been evident in allocating gas supplies, the BGC cannot have lowered the price of gas to consumers in order to pass on the surplus. An alternative route to enlarge the consumer surplus is via income transfers through general taxation. For this route to be operative the BGC must relinquish its profits to the Treasury. Since no such relinquishments have been made – at least until recent times when legislation forced the BGC to pass on an excess profits levy – it is doubtful that this route has had any significant impact on the redistribution of economic rent. As a corollary to this reasoning, much of the BGC's surplus must have remained in the BGC, absorbed in terms of organisational slack.

In effect, the monopsony will have transferred economic rent from large private oil companies to a large public corporation, the BGC. In the process, the consumer surplus will probably have been reduced compared to an unregulated market since fewer fields will have been developed and recoverable gas reserves will have declined.

Collecting Oil Profits

Prior to a detailed analysis of the impact of North Sea oil taxation in Chapter 3, we first outline some of the chief drawbacks of the taxation system as an instrument in collecting oil–field profits.

Taxing Revenues

With a given capacity, the effect of a fixed rate of revenue tax will be to lower the company's marginal revenue schedules by a fixed percentage. The relative marginal incentives directing output decisions would not be affected but the incentives to install capacity would be affected. Marginal net operating profits would decline, bringing the marginal benefit schedule down by a constant and absolute amount and thus reducing the attractiveness of capacity installation. By making capacity less attractive, recoverable reserves will be lower with the bulk of the decline in production coming after peak supplies have been reached.

Taxing Profits and the Normal Return on Capital

If profits are defined in a way that does not coincide with pure profits, then production and investment decisions may be affected. Since the system of North Sea oil taxation has no allowance systematically linked to the length of delay before a field comes onstream or the length of payback, it cannot hope to tax pure profits and only pure profits. In some circumstances the allowances related to capital expenditure may exceed the normal return to capital; more often, though, the capital allowances underestimate the return to capital.

The tax system then represents a tax on capital and will be a disincentive to installing North Sea capacity.

Taxing Profits and Risk Bearing

In the margin, expected profits must be positive to attract investors to bear risks. The lack of any distortion rests on leaving the marginal profits unaffected and assumes the marginal risk premium will not be altered. Companies will have different risk premia and it seems unlikely that any tax system will have a neutral impact on risk profits for the wide range of North Sea projects and companies. Given the imperfections in the capital and risk markets, distortions because of the tax system are inevitable. Indeed, given these imperfections, distortions that tend to increase the extent of installed capacity or delay output may be desirable. For North Sea taxation to offset distortions caused by CT and risk profits it must transfer the financial risks to the government more than proportionately to the transfer of expected return. To achieve this, taxes as a proportion of profits must decline as capacity increases so that tax take declines with marginal profits. Since such progressiveness is lacking from the North Sea tax system the system of taxation aggravates the capacity and output misallocations.

Distribution of Profits

Given the international nature of the world oil market, the consequence of the tax system depressing North Sea capacity and bringing production forward will have negligible impact on the price paid by the consumer. The consumer surplus will not be affected by North Sea profits taxation and only the producer surplus need be considered here. Whether this method of profit collection would yield greater revenues for the Treasury than the licence bids made in the unregulated market will depend on the extent of competition in the industry.

2.6 CONCLUSION

Although government oil and gas policies tend to have commendable intentions, the impacts of policy are rarely desirable and the occasions when policy acts in the opposite direction to that intended are not infrequent. Furthermore, unintended interactions between different targets imply that – even when a particular policy comes close to achieving its intended aims – the policy, aggravates other difficulties.

This chapter has sketched the background for a more detailed analysis, and has also framed the terms of reference. It has determined both the objective which is furthest from being achieved and the policy which operates as the greatest hindrance. The final column in Table 2.2 suggests the objective which is most frequently hindered by policy is one of two objectives ommited from the list of intentions recognised by government: namely, the overall development and production of North Sea reserves (as distinct from the timing of development or production).

In this area policy appears to aggravate the misallocation of resources more severely than in other areas. Moreover, recognition of this area as problematic, let alone recognition of the aggravation of the problem by policy, is not common. Oil taxation appears to be the policy instrument whose impact is furthest from its intentions and when the detrimental effect on development decisions is likely to be greatest. These findings are explored in detail in subsequent chapters.

Tax policy is not one element of an integrated package towards the North Sea exploration and production sector. It is thus appropriate to adopt a pragmatic analysis of the effects of taxation, without undue consideration of the interactions between tax policy and the other elements of policy.

Tax Policy and Oil Profits

One consequence of the government's decision not to auction licences was that the economic rent associated with oil–field production would, by and large, be passed on to the private sector. For the government to participate in the rent then, it adopted a tax policy that between 1979/80 and 1983/4 collected nearly £30 billion out of an estimated total valued added for the years 1979–83 of over £56 billion (billion is thousand million throughout the book). The tax policy, however, does not just alter the distribution of profits between the government and the private sector but also alters the magnitude of profits by interfering with development decisions. Indeed, there appear to be five chief problem areas in the way current loss policy is used to collect oil–field profits, each of which imply tax policy has implications for efficiency. For the moment we merely assert that each problem exists and outline the efficiency implications. Evidence for these assertions and a more detailed account of their consequences is discussed in Chapter 3.

Absence of Progressivity

As the tax take fails to rise progressively with profits, the current tax system will tend to exhibit similar features to those of a tax on revenues. In terms of Figure 2.1, the effect of a tax on revenues will be to lower the marginal revenue schedules – by a fixed percentage if the rate of revenue tax is fixed – and hence reduce period-by-period marginal net operating profits (Figure 2.2). This would cause the marginal benefit schedule of Figure 2.4 to decline, the attractiveness of capacity installation would fall and commercially recoverable reserves will be reduced.

The Return on Capital

If profits for tax purposes are defined in a way that does not coincide with pure profits, then the return on capital in the margin may be reduced. Since capital allowances in North Sea oil taxation are not systematically linked to the length of delay before a field comes onstream or the cost of capital, North Sea taxation cannot hope to tax pure profits and only pure profits. In some circumstances the allowances related to capital expenditure may exceed the normal return to capital; more often, though, the capital allowances will underestimate the return to capital. The tax system then represents a tax on capital and will be a disincentive to installing North Sea capacity. In terms of Figure 2.5, a tax on pure profits will not only reduce the marginal benefits in proportion to the rate of tax but, by allowing fully for the economic cost in installing capacity, will also reduce the marginal capacity cost schedule by an identical proportion. If the normal return on capital is not protected, however, the marginal capacity cost schedule will not drop by as much as the fall in marginal benefits. Consequently less North Sea capacity will be installed and recoverable reserves will be lower because of the tax system.

Risk

Any tax on profits will have two implications for risks. First the expected level of profits will be reduced and, since one component of expected profits is the return necessary for bearing risks, the incentives for risk-taking – that is, for installing capacity – will be moderated. Second, any profits-based tax system will alter the scale of risks freed by the company, not only by reducing the possible rewards but also by limiting the extent of possible losses. That is, providing some degree of tax offset is available, a tax on profits will also act as a subsidy to losses. The first implication will reduce the risk premia that may be earned on North Sea investments, while the second implication will affect the risk premia required by companies if they are to be exposed to the possibility of making losses. Only in the extremely unlikely event that required risk premia are reduced by exactly the same amount as available risk premia will the tax system not affect development decisions. As a second-best criterion, for a tax system not to deter development decisions the financial risks in oil–field developments must be transferred to the government more than proportionally to the transfer of expected return. To achieve this, taxes as a proportion of profits must decline as capacity increases and the tax take must become increasingly less significant as profitability deteriorates. Since such progressiveness is lacking from the North Sea tax system, the system of taxation aggravates the capacity misallocations caused by risks.

Taxing Incentives

Governments have shown they are keen to raise tax rates or reduce tax concessions whenever oil prices and North Sea oil profitably increase. By establishing a link between taxation and profitability, however, the government will act as if it is taxing incentives, and the private sector will feel it is facing a one-sided future where taxes will rise if real oil prices increase but will not fall symmetrically if real oil prices drop. Hence the desirable unregulated market reaction to higher real oil prices – that is, to increase oil supplies – will be blunted.

Uncertainty

Although a significant move in oil prices seems to be sufficient for governments to adjust North Sea taxation, it is not necessary and governments seem likely to adjust North Sea taxation whenever it is politically expedient. This adds further uncertainty to oil–field economics which will deter development decisions in two ways: first, by creating uncertainty over when tax changes will occur, and second, given the complexity of North Sea taxation, by creating uncertainty over what tax changes will occur.

Each of these five problem areas suggests North Sea oil taxation will deter development decisions and hence the scale of profits from North Sea activity. North Sea oil taxation, however, was designed to affect the *distribution* of profits only. Given the international nature of the oil market, the depressing effect on North Sea capacity installation will have a negligible impact on the price of oil paid by consumers, and thus the consumer surplus will not be significantly altered. The producer surplus, though, will be altered. Whether this method of rent collection yields greater revenues for the Treasury than the licence bids that would have been made in an unregulated market will depend on the extent of competition in the industry.

3 The Economics of Oil Taxation

A detailed analysis of the main economic problems associated with the offshore oil taxation system is presented in this chapter. Five problem areas of the current oil taxation policy are considered: namely the absence of progressivity, the return on capital, risk, taxing incentives and uncertainty. For each of these five areas, we first define the problem and then outline the relevance for North Sea oil development decisions. Evidence of the problem's existence in the current tax system is then discussed, and finally the lessons to be learnt if the system is to be improved are highlighted.

3.1 ABSENCE OF PROGRESSIVITY

Definition

Progressivity in a tax system implies that as profits increase the tax take, as a proportion of pure profits, increases at a more than proportionate rate. Pure profits may be defined as those profits in excess of the level necessary to induce economic activity. In practice, when pure profits are low, a progressive tax system will collect negligible sums, but when pure profits are substantial a significant proportion will be colected in tax. Progressivity should be an automatic feature of the tax system so that the tax take should increase progressively without changes being made; especially structural changes, for instance, to the allowances. If progressivity is not automatic then changes are made on an *ad hoc* basis and instability is likely to become a problem.

In the offshore oil industry, pure profits are the residual sum above normal profits which will include some return on capital and on risk, (see sections 3.2 and 3.3). As prices or reserves increase, costs decrease or production is brought forward (that is, as factors affecting pure profits alter), then the percentage increase in tax should be greater than the percentage increase in pure profits.

A major constraint on the mechanics of a progressive tax is the need to avoid 'gold-plating'. 'Gold-plating' arises when marginal real discounted profits rise with unproductive capital expenditure and occurs when a field's marginal tax take exceeds 100 per cent. In such circumstances, spending will reduce the payments by more than the level of spending and the normal process of profit maximisation will result in wasteful, unproductive expenditure.

Relevance to Development Decisions

Conceptually, it is not relevant to development decisions whether a tax is progressive or not as long as the tax base is pure profits. However, problems arise in attempting to identify pure profits, particularly in assessing the necessary returns on holding risks. In the absence of risks, profits would be defined as the NPV deflated to account for inflation and discounted at a rate to reflect the cost of money: for example, currently around 10 per cent per annum. This measure would indicate the total profitability of a project. Since our concern is about the profitability of one project relative to another, and since companies are likely to be indifferent between owning a large percentage interest in a small discovery and a small percentage interest in a large discovery, this measure needs to be scaled according to the size of the project. That is, rather than considering total NPV, the relative profitability is better shown by the NPV per unit of output (for example, per tonne). On this measure differences between profits and pure profits are likely to be a fixed item. This is because the cost of bearing risks per unit are not (by and large) project specific (that is, they tend to relate to the exogenous oil price), nor will they increase with field size. Hence the tax needs to be progressive with respect to this measure of profits if it is not to tax returns on holding risks.

When compared to a non-progressive system, a progressive tax system has three beneficial implications for development decisions. First, for a given tax revenue target a progressive tax system will be less likely to impair project economics to such an extent that development decisions are jeopardised. The neutrality of a progressive tax system may be illustrated by considering a competitive auction for UKCS tracts.

In a competitive auction for UKCS licences, companies wishing to exploit a prospect would be forced to bid up the price of the tract by the fear of not securing the prospect, and the supplier (in this case, the British government) would tend to receive all the profit over and above the costs of exploration and extraction. Hence the pure profits or economic rent associated with non-renewable resources would be collected by the government. In this auctioning process, the willingness of companies to pay for a tract's mineral rights will tend to increase more than proportionally with the expected NPV of the tract because required risk premia will become less significant. That is, there is some theoretical level of expected value below which the profits will not cover the costs of bearing risks, and above this fixed element pure profits will be a progressively increasing proportion of the NPV. Since the amount companies bid will not affect their eagerness to develop a prospect, progressive revenue collection will not, in principle, distort development decisions.

In practice, UK governments have favoured a discretionary system of licensing which involves negligible rental payments for the mineral rights, especially for fields granted development consent after April 1982. For those areas granted development approval before April 1982, a fixed rate of royalty is charged in the gross value of oil produced in lieu of payment for the mineral rights. Generally, royalties represent a great underpricing of a tract's value, and profits are passed to the oil-producing companies and away from the government supplier of mineral rights. In these circumstances the surplus will be received by oil companies and will rise progressively with NPV. Since companies would have been willing to bid this surplus in an auction system and would still have been willing to develop any discoveries, this surplus may be taxed without distorting marginal development decisions. Therefore by mimicking the outcome of a competitive auction a progressive tax system will not affect development decision-making. At the point where companies are just earning marginal profits, the impact of a progressive system will be negligible compared to a non-progressive system, and the tax will be less likely to be a burden on risk profits.

Second, if the tax is progressive then the detrimental impact of other imperfections in the tax system will not be magnified. For example, if the tax base includes part of normal profits then there will be a reduction in marginal field profitability. If the tax system is not progressive then this reduction may be severe and development decisions on many fields may be jeopardised. If, however, the burden of the tax system on less profitable fields is small (that is, if the tax is progressive) then the reduction in marginal field profitability caused by the tax base, even though it includes an amount of normal profits, will be less significant. This reasoning applies to all imperfections in the tax base. Furthermore, if the tax system is progressive, then it is more likely to be stable. Under a progressive system price rises would automatically cause tax revenues to increase and there would be no need to change the system's rates or allowances to bring additional revenues.

The third beneficial implication of a progressive tax system is that signals to the developers of marginal fields will not be dampened. When prices fall or inflation rises, in a progressive system the impact on post-tax profits will be cushioned for the more profitable fields, but will not be cushioned for fields which are less attractive. Therefore when the message of a price or cost signal is pertinent (that is, when a discovery is on the verge of being developed) a progressive tax system will in no way dampen the signal to companies and instead will only tax those fields well above the level of commercial viability. Hence, when a field is just earning enough profits to warrant development, taxes will not be relevant to incremental economics and will not interfere with development decisions.

Concern exists in the oil industry that as a consequence of progressiveness the incentives to develop incremental barrels on very profitable fields are dampened. This is because under a progressive system the tax rate on developing an additional barrel from a profitable field will be relatively high. Although this criticism is valid, this effect is the lesser of two evils. It seems preferable to extract an incremental barrel from a profitable field only if the tax system avoids making less attractive projects uncommercial.

Apart from these three beneficial implications for development decisions, a progressive tax system would also adhere to the equity principle that 'the rich should pay more'.

Evidence

Progressivity may be implemented either by having a tax base that is proportional to profits with a rising schedule of tax rates, or by having a single tax rate with a tax base that grows progressively with profits. Of course, these two methods may be combined.

In the UK, the government opted for a tax system without rising schedules of tax rates. To discover whether the tax system is progressive or not, it is therefore necessary to find out whether the tax base rises progressively. Two forms of progressivity need to be considered: first, progressivity with respect to inter-field profitability and second, with respect to prices (or costs) on any given field.

Petroleum Revenue Tax: Progressivity Across Fields

Within the Petroleum Revenue Tax (PRT) tax base, the progressive element is supposedly provided by the oil allowance, which was introduced in 1975 'after forceful representations by the oil industry'.[1] The oil allowance operated per field, and initially exempted from PRT 1m long tons per year up to a cumulative total per field of 10m long tons. Although the size of the oil allowance has subsequently been altered twice, its aim remains to provide a progressive incidence in the tax system and to aid less profitable discoveries.

Being calculated purely in terms of a fixed quantity of oil, the oil allowance is of greater benefit to small fields than to large fields. If small fields are consistently less profitable than large fields then the oil allowance would be progressive. It is a fallacy, however, that big fields are necessarily more profitable per unit than small ones. For instance, both Magnus and Brae are large but marginal oil fields whilst Argyll and Auk are small and amongst the most profitable projects. In practice, the advantage of the oil allowance amongst the established commercial projects (where our field information is more reliable) is enjoyed more by profitable than by less profitable fields. Table 3.1 demonstrates this assertion. The two rewarding but short-life projects, Argyll and Auk, and the middle-ranking (in terms of real profitability) Montrose, enjoy significant increases in their profitability both in terms of rates of return and NPV per tonne, as a result of rates of the oil allowance. Furthermore, the oil allowance is a great help to Dunlin and Murchison UK, two of the most promising fields on the UKCS. On the other hand, some of the discoveries which the allowance was designed to help are offered negligible aid (for example, Cormorant North, Alwyn North and Magnus).

Similar anomalies exist with respect to new oil–field projects (see Table 3.2). Marginal but medium-sized Bruce is helped relatively little by the oil allowance. However, the small but profitable Glamis and Emerald benefit significantly.

By comparing the tax takes with and without the oil allowance, its effect on the progressivity of the system across fields can be ascertained (see column (b) of Table 3.1 for the established commercial fields and column (b) of Table 3.2 for potentially commercial fields).

Table 3.1 The impact of the oil allowance on commercial oil fields ranked in order of their pre-tax profitability (as measured by NPV per unit of production)

	Advantage of oil allowance	
	(a) to real IRR (%)	*(b) to real NPV (£/tonne)*
Piper	1.65	1.89
Forties	0.62	0.75
Argyll	9.36	18.41
Murchison UK	2.6	3.46
Claymore	1.53	2.36
Auk	7.25	13.2
Dunlin	2.7	3.94
Fulmar	2.36	2.44
Ninian	0.54	0.76
Thistle	1.48	2.18
Beryl	2.13	1.89
Brent	0.16	0.26
Buchan	4.35	6.67
Brae	1.82	2.39
Duncan	19.84	16.81
Stratfjord UK	1.0	1.39
Cormorant North	0.52	0.87
Cormorant	6.3	12.81
Maureen	2.75	4.73
Heather	3.36	6.62
Montrose	4.66	10.25
Magnus	0.80	1.07
Hutton NW	3.14	4.46
Beatrice	1.92	3.55
Balmoral	13.53	12.29
Brae North	7.69	6.64
Hutton	1.43	2.09
Alwyn North	1.29	1.88
Highlander	11.02	7.59
Clyde	2.89	3.91
Tartan	1.45	2.03

IRR = Internal Rate of Return.

The range of the tax takes across all fields in production or under development is widened by incorporating the oil allowance. However, the fields of lowest profitability are not the ones to receive the greatest benefits and the impact of the allowance on fields of similar profitability varies widely. This anomalous effect of the oil allowance

Table 3.2 The impact of the oil allowance on potentially commercial fields ranked in order of their pre-tax profitability (as measured by NPV per unit of production)

	Advantage of oil allowance	
	(a) to real IRR (%)	*(b) to real NPV (£/tonne)*
030/17	11.25	1.21
Glamis	6.44	8.79
Mabel	7.69	7.1
Ettrick	6.93	8.18
Emerald	5.67	5.91
Columba	8.23	5.26
T-Block	2.23	1.57
Kittiwake	7.45	6.45
Montrose South	7.89	6.9
Scapa	12.18	7.24
Crawford	4.09	2.2
Claymore North	11.04	7.25
Lyell	2.31	1.77
Renee	7.69	4.34
Andrew	5.91	4.72
Tern	3.39	2.86
Don	2.28	2.2
Eider	3.04	2.6
003/28	7.56	3.1
Galley	3.09	1.71
Clair	3.72	2.65
Alwyn South	3.01	1.84
030/13	2.47	1.21
Bruce	1.33	1.27

IRR = Internal Rate of Return.

is due largely to the fundamental and mistaken premise that small fields are marginal fields.

Progressivity as Profitability Alters

In a progressive tax system average tax takes increase as profitability increases; thus for any field its marginal tax take must be higher than its average take. Indeed, if the tax system works well, the marginal tax takes across a variety of fields will decline towards the maximum tax rate as the average tax takes increase towards the maximum tax rate (Figure 3.1).

Figure 3.1

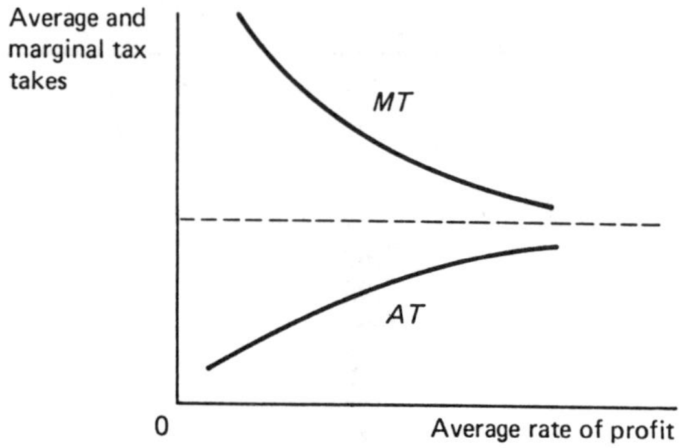

MT = Marginal Take
AT = Average Take

It is worth drawing attention to the number of fields under the existing tax system where marginal government takes are less than average government takes (Figure 3.2). This indicates that as prices (and profitability) rise, average tax takes will fall. Some nine of the established fields (Auk, Murchison UK, Fulmar, Montrose, Buchan, Tartan, Dunlin, Heather and Cormorant), have lower marginal takes than their average tax takes, which indicates a tax system poorly related to profitability.

CT

CT was not designed to be progressive. The CT base comprises net revenue less interest on debt, less accrued royalties, less a deduction for capital spending and less a deduction for PRT payments, and in itself regressive. First, although pure profits are reduced by risk premia and normal returns to capital, the CT base is unaffected by these changes: an effect accentuated because PRT is non-progressive and PRT is a CT allowance. Second, in assessing the ability of CT to become less onerous when oil prices decline and risk premia become relatively more important, we have found that without exception, CT collects proportionately more from each established field when prices are lower (and the risk premia is more necessary). For some fields (notably Cormorant and Heather where pre-tax unit profits are amongst the lowest) this effect is most significant.

Analysing the impact of CT across a selection of potentially com-

Figure 3.2 Average and marginal tax takes plotted against pre-tax rate of return under the 1984 tax system

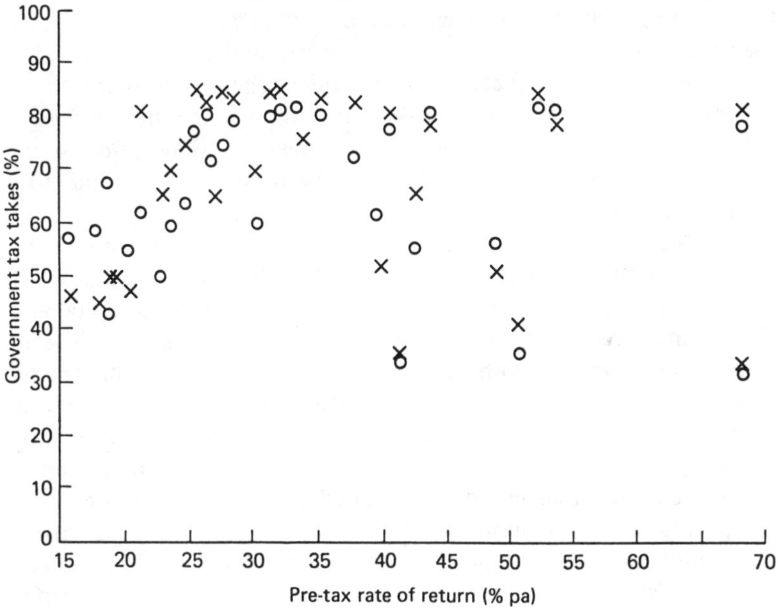

O = Average tax take
K = Marginal tax take

mercial fields illustrates the way CT behaves in a non-progressive fashion. Relatively unprofitable new fields such as Clair, Don and Tern receive a proportionately greater CT burden on unit profits than do the more profitable Andrew and Mabel.

Lessons

Not only does the tax regime fail to collect the economic rent in a progressive manner by isolating the most prosperous fields from the less prosperous projects, but it fails to be progressive for individual field projects. That is, the tax system is neither progressive inter-field nor progressive intra-field.

PRT is regressive because the PRT base does not grow progressively with profits. In a tax system where the tax base is equal to pre-tax revenues then, with a single tax rate, the average tax take would equal the marginal tax take. If the tax calculations included allowances that reduced the tax base below pre-tax net revenues, then the marginal tax take would be higher than the average tax take. Conversely, if the tax base were greater than pre-tax net revenues then the marginal tax take would be lower than the average tax take. For those old fields where royalties collect the major share of tax revenues, this latter possibility is clearly likely since the tax base is closer to gross revenues than to net revenues. This is also the case for new fields demonstrated by the regressive nature of PRT and, to a lesser extent, of CT.

For a tax system to be progressively related to profitability in the absence of a rising schedule of tax rates, it must have a tax base that is lower than net revenues to reflect both a normal return to capital and risk premia. We show in subsequent sections how difficult it is to quantify and measure both these factors. Here, we found that the oil allowance is an unsatisfactory concession which does not distinguish between less profitable and more profitable projects in its tax-reducing impact. Furthermore, we doubt if any concession or combination of concessions could be designed to distinguish between the less profitable and the more profitable projects given the wide variety and complexity of economic circumstances surrounding North Sea oil fields. Consequently we are sceptical that a tax base can be designed that will rise progressively with profits.

Instead, the principal lesson of this section is that progressivity should be implemented via a rising schedule of tax rates, with a tax base that is proportional to profits.

3.2 THE RETURN ON CAPITAL

Definition

It is emphasised throughout this chapter that for a non-distortionary tax the tax base should be pure profits. If a tax system is to be based on pure profits (aside from the considerations for risk – see below), then its cost allowances must reflect not only the scale of expenditure but also the timing of expenditure relative to revenues.

Pure profits are the net operating profits over and above the level of profits reflecting the normal return on capital. The normal return on capital reflects the rate of return available on the most profitable alternative use for that capital, a use which is crowded out by pursuing this activity. In the North Sea, the normal return on capital is the level of profits just sufficient to prevent capital from moving away from the North Sea into other uses and will determine a company's discount rate. The greater are the costs or the longer are the lead times then, *ceteris paribus*, the higher will be the absolute level of profit that needs to be set aside as the normal return on capital.

Relevance to Development Decisions

When the tax system represents a tax on capital, there will be a disincentive to installing North Sea capacity. That the tax system may act as a tax on capital is of particular importance in the North Sea oil industry because of the industry's capital intensity. In tax systems that do not have full allowance for lead times and heavy front-end loading of capital expenditure, the normal return on capital will tend to be depressed. This incidence may, in part, be passed on to other economic agents. If it can be passed on then the economic consequences in long production cycle industries, such as North Sea oil, will be dissipated. There are three directions in which the tax on capital may be shifted.

First, if the supply of capital is price inelastic then any burden of taxation may be passed from the users of capital on to the suppliers of capital. By depressing the marginal return on capital the demand for capital will be reduced and, with inelastic capital supply, the rate of interest on capital will fall. In the UK oil extraction sector, though, this outcome is not likely. Even where the total supply of capital may be very inelastic, the supply of capital to the UK oil industry is likely to be highly elastic given the sophisticated nature of the UK financial markets.[2]

Second, a tax burden on the users of capital may, in part, be passed on to the consumers of the final product. This will be possible if the price elasticity of demand for the final product is low. Demand for UK crude oil, however, is very elastic. UK producers tend to act as price-takers, which reflects the ease of substitution by consumers of UK and non-UK crude oil.

The third direction to pass a tax burden on capital is towards other factors of production, in particular towards labour. A tax on capital will alter the relative attractiveness of labour such that more labour-intensive production techniques will be adopted. This effect is a substitution effect, but it is not the only effect. There is also an output effect – since costs will have risen, the scale of production will tend to fall – which will be likely to reduce the demand for labour. In situations where the output effect tends to dominate the substitution effect, the demand for labour will fall. If the supply of labour is price inelastic, then the wage rate will decline and thus the tax burden on capital may be passed on to labour. However, this outcome is not likely in the UK oil extraction sector where the substitution effect will tend to dominate the output effect, and hence the demand for labour may not fall. Moreover, even if the demand for labour falls, the labour supply in the UK oil extraction sector represents a skilled and mobile workforce, and will tend to be elastic with respect to the wage rate, so it will not be possible to pass any tax burden on the return to capital on to other factors of production.

Any tax burden, therefore, must be faced by the users of capital (that is, the oil companies) and, by and large, cannot be transferred into lower returns to shareholders, into higher final product prices or into lower wage rates.

Evidence

Does the PRT Uplift Provision Protect the Normal Return to Capital?

In the absence of any special provision for the normal returns to capital, the return to capital employed on the UKCS would be reduced compared to the return received by other resources offshore and compared to the return on capital in other industries because of the long lead terms. Capital would be diverted away from the North Sea into less productive spheres. While this misallocation of resources is likely to be greater in absolute terms for capital-intensive projects, it will be most essential for the projects with the longest lead times where the importance for the tax to be assessed on real discounted profits rather than nominal undiscounted profits is greatest.

The uplift provision in PRT is the feature in the North Sea tax system most likely to alleviate this distortion. It creates an additional tax relief for development costs (relative to tax relief on operating costs) which tends to reduce any intra-industry distortion. However, examining the uplift's impact shows the inter-industry disincentive is not rectified.

In estimating the advantage of uplift on fields already granted development consent (that is, before April 1982), we find that although some projects gain little, others such as Thistle, Ninian, Piper and Claymore enjoy great benefits both to their real internal rates of return and to their NPVs per tonne. In general as project lead times shorten, the advantage of the uplift increases. Indeed most of the projects with the longest lead times, which therefore are likely to need greater protection for the normal return on capital, receive little benefit from the uplift provision. Conversely oil fields that are less likely to be seriously affected, because lead times are shorter, find the uplift valuable.

With respect to oil fields yet to be granted development approval, the impact of the uplift provision generally seems arbitrary between fields of varying lead times (Table 3.3 below). The fields with the

Table 3.3 Impact of PRT uplift provision on potentially commercial oil fields ranked according to length of lead time

	Advantage of uplift provision	
	(a) on real IRR (%)	*(b) on real NPV (£/tonne) (10% dis.)*
Clair	0.58	0.43
Galley	0.84	0.4
Lyell	0.92	0.54
Tern	0.56	0.4
Crawford	0.71	0.34
Eider	0.92	0.65
Alwyn South	0.79	0.38
Bruce	0.4	0.39
Ettrick	0.52	0.5
Mabel	0.82	0.66
Renee	1.18	0.34
T-Block	2.1	0.86
Claymore North	0.48	0.34
Columba	0	0
Kittiwake	1.58	0.56
Kittiwake	0.21	0.17
Montrose South	0	0
003/28	0.71	0.21
030/7	1.73	0.9
030/13	0	0
Don	0	0
Emerald	0	0
Glamis	0	0
Scapa	0	0

shortest lead times, which least warrant uplift, tend to be small fields with production shielded from PRT by the oil allowance and hence, desirably, receive little or no benefit from the PRT uplift allowance (Don, Glamis, Emerald, Scapa, Montrose South, Claymore North). However, apart from these very short lead time fields, uplift operates rather arbitrarily and there are a number of fields with similar lead times which enjoy very different boosts to profitability as a result of uplift. Most worryingly, though, there are some projects, such as Tern, which have relatively long lead times but are given relatively little assistance as a result of the uplift clause.

It seems that PRT introduces additional distortions and, in many cases, aggravates the diversion of capital towards less profitable non-North Sea activities by failing to protect the return on capital.

CT

Only if the CT base is identical to pure profits will the tax be non-distortionary. In the margin, post-tax profits will be zero where pre-tax profits are zero and maximising post-tax profits would then yield identical decisions to maximising pre-tax profits. CT, however, is a tax on the excess of revenues over historic costs, and may affect development decisions since historic costs are a poor reflection of normal returns on capital. Indeed, as is apparent in this chapter, only under rather rare circumstances will CT be a neutral tax (see Chapter 2, section 2.2). Moreover, since the oil extraction sector displays economic characteristics which restrict the opportunities for passing on the burden of the tax, the direct impact cannot be dissipated. In the margin, pre-tax discounted profits must be positive, and pre-tax rates of profit must lie above interest rates, for post-tax discounted profits not to be negative. Consequently, investment opportunities will appear depressed, investment will be held back, capital equipment will be underutilised and fewer North Sea oil fields will be developed than might be developed in the absence of CT.

The impact of CT on unit NPVs shows that some of the prospective new developments with the longest lead times (Crawford, Mabel, Bruce) have their profits significantly reduced by CT (see Table 3.4).

Lessons

Because capital costs in the tax system are assessed in historic terms, the tax incidence tends to fall not only on pure profits but partly on

Table 3.4 The impact of CT on potentially commercial oil fields as ranked by length of lead time

	Reduction in real IRR as a result of CT (%)	*Reduction in real NPV as a result of CT (£/tonne) (10% dis.)*
Clair	3.45	3.44
Galley	4.01	2.79
Lyell	4.2	3.53
Tern	4.26	4.36
Crawford	4.55	3.13
Eider	4.64	4.65
Alwyn South	4.73	3.27
Bruce	3.98	4.64
Ettrick	4.52	7.8
Mabel	6.13	7.74
Renee	6.23	4.3
T-Block	5.44	3.8
Andrew	5.55	5.69
Claymore North	7.45	7.34
Columba	6.74	5.42
Kittiwake	6.05	7.18
Montrose South	6.11	7.59
003/28	7.81	3.75
030/7	8.7	7.86
030/13	6.2	3.59
Don	6.29	5.22
Emerald	7.07	8.31
Glamis	5.13	9.66
Scapa	7.51	6.39

the normal return to capital. Even the PRT uplift is just an arbitrary multiple again based on historic costs. For the tax system to leave the normal return to capital unaffected, an allowance reflecting the normal return to capital must be incorporated in the tax system. There are two components to this allowance which are absent: first, the real rate of interest that is available in alternative investment opportunities, and second, an item to reflect inflation between the date of expenditure and the date of allowance against tax. The second component is necessary to ensure that tax is assessed on real profits rather than on nominal profits, and requires that depreciation should occur on replacement cost rather than on historic cost values. Long production cycle industries such as North Sea oil are adversely affected by the absence of both these components.

We have demonstrated that the uplift allowance in PRT is too rigid to reflect these two components. Even if it could CT would still have a detrimental impact, particularly as a consequence of the limited first year allowances for capital expenditure. Thus an allowance for capital costs is needed which not only reflects the level of spending but directly and automatically increases as lead times lengthen, as inflation rates alter, and as interest rates change.

3.3 RISK

Definition

Exploration activity is dominated by risks; risks of drilling dry holes, risks of delineating a poorly defined deposit and risks in evaluating the commercial prospects of a discovery. Generally, in the exploration stage, the risks are related to the unknown about the geophysical attributes of a tract, whilst financial risks are more pronounced once development is underway. In the development stage of an oil field project, the heavy front-end loading of capital costs exaggerates the significance of the risks in offshore oil production and development decisions.

Relevance to Development Decisions

In the absence of a complete set of forward and state contingent markets, oil producers must base their decisions on expectations of future demand. As demand in a future period will only be reached through a sequence of spot markets a risk of misestimating demand must be borne by companies developing oil fields that cannot be insured against by hedging in state contingent markets. With a range of possible demands for a given period, companies will be inclined to avoid such risks. Even assuming expectations are rational and the central estimate is unbiased, the wider the range then the more production will be shifted out of a period associated with greater risks and either into other periods or never produced. As more distant periods are associated with greater uncertainty, the effect of risks tends to be to generate a more rapid rate of depletion from a smaller recoverable reserve base. That is, an identical expected value is less attractive than a known certain cost. Hence, the existence of financial

Figure 3.3

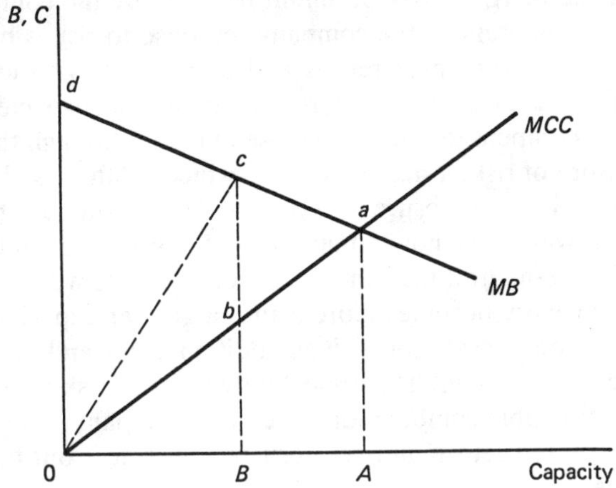

risks reduces the incentive to delay production and makes capacity installation less attractive.

The Relevance of Risk Profits to Capacity Decision

In the margin, expected profits from installing North Sea capacity must be positive to attract investors to bear risks. In Figure 3.3 the marginal benefit schedule represents expected marginal benefits and the marginal capacity cost schedule is costs excluding risk premia. In the absence of taxation the capacity choice would be less than *A* at, say, *B* where the positive marginal profits (*cb*) are just sufficient to compensate for bearing risks.

A tax system based on pure profits would leave the marginal units of capacity at *B* untaxed. Because of risks, the producer surplus would be less than *Obcd*. The risks reflect the danger that the company will make a loss by investing in this project, that is, earn a lower return than available elsewhere. The size of the necessary risk premium will then tend to be proportional to the scale of capital expenditure. Hence the producer surplus might be *Ocd*, which if adopted as a tax base would probably not be distortionary.

Such a tax system is not, however, guaranteed to be non-distortionary. The lack of any distortion rests on leaving the marginal profits unaffected and assumes the marginal risk premium will not be altered. However, risk premia are likely to vary with a company's total profits: hence, taxing intra–marginal profits will tend to affect the size of the marginal risk premium required by the company.

One way of expressing the company response to risk is by considering their risk-return preferences and opportunities.[3] Indifference curve analysis measures the degree of additional expected return necessary to compensate for an increase in risk. Although there is no single measure of risk, a simple measure which is often used, and will be used here, is the exposure to a possible loss. Families of indifference curves will show how company preferences are ranked. If a company is risk-neutral then the indifference curves will be horizontal; as risk-aversity becomes more acute, a greater expected yield is necessary to compensate for a given additional risk and the indifference curves will steepen. The opportunity locus of risk-return points shows the attainable combinations open to a company. The trade-off between expected return and greater risk thus traces out the opportunity locus.

In Figure 3.4 beyond point A' (and B') greater risks will be taken but expected total returns will decline because marginal expected profits are now negative: that is, A' corresponds to the peak of the locus. As points beyond A' involve greater risks and lower expected returns, and points before A involve lower expected returns but no less risk, a pre-tax profits-maximising risk-averse company will never choose a risk-return combination outside the segment of the opportunity locus between A and A'. Given that risks are directly proportional to capacity, the slope of this segment reflects the size of the expected marginal capacity net profits. The less sensitive marginal profits are to capacity then the flatter will be the opportunity locus. The company's optimum position will be the attainable combination which maximises corporate utility; that is, where the indifference curve is just tangential to the opportunity locus (point a). Clearly the consequences of risk for North Sea capacity may be measured by the distance from A', which would be the point reached either by a risk-neutral company (horizontal indifference curve) or by a risk-averse company facing full and fair insurance possibilities (horizontal opportunity locus through A').

In the risk-return analysis of Figure 3.4, a tax system that does not affect marginal trade-offs between risks and returns would move the opportunity locus from AA' to BB' where the ray from the origin abO shows the same level of development associated with different levels of taxation. Only if the slope of the company's indifference curve at b is the same as the slope of the opportunity locus at b will

Figure 3.4

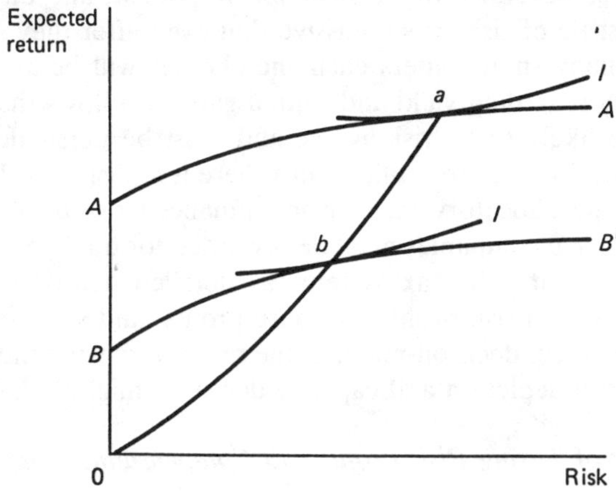

*ab*O ◊ Same development with different tax
A' Pre-tax opportunity locus
B' Post-tax opportunity locus
I ◊ Indifference curve

the tax system be non-distortionary. Since there are no *a priori* reasons for this identical slope – and indeed the move from *a* to *b* will differ according to each company's circumstances – it is extremely difficult for a tax system to have no effect on capacity installation.

For a tax system to have no impact on the capacity choice it would have to reduce the opportunity locus in a way which imitates each company's indifference curves. Although it is possible to envisage a tax system that might be capable of this imitation for one or two companies, it is not possible to conceive of a system capable of such neutrality for the wide range of North Sea projects and companies. (We could repeat these comments over the choice between relatively certain profits from early production and relatively uncertain profits from later production to suggest any system is also likely to affect a field's rate of depletion.)

The Role of Risk Profits in Capacity Decisions

Not only are the risks greater in the offshore industry than elsewhere, but as insurance possibilities are strictly limited they are economically more significant. On the one hand the risks in major variables such as oil prices or cost inflation tend to affect all projects in the same direction, so several prospects cannot be pooled, and on the other hand the scale of risks is so massive that even after they are spread amongst many shareholders each shareholder will be aware of the uncertainty regarding yield and capital gain. It follows that oil companies are likely to be risk-averse and must be persuaded to bear risks via receipt of pure profits. Thus there is a positive role for pure profits in the allocatory mechanism. Financial risk-bearing implies that risk-averse companies must be rewarded for holding risks by the receipt of profit. The tax system (as detailed below) cannot distinguish between risk profits and pure profits, and will tax both. As risk profits affect decision-making, the tax system will tend not to be neutral: both depletion and capacity decisions might be affected.

The Effect of Taxing Risk Profits and Compensating Losses

Taxation changes the risks, not only reducing possible gains but also, because losses are allowable against tax payments elsewhere, by returning possible losses. Risks will tend to deter capital installation as producers will not wish to be exposed to the uncertainty inherent in the value of oil production. A tax system will scale down the variance in marginal benefits from capacity installation. If the post-tax capacity choice is unchanged compared to the pre-tax, the reduced variance in marginal benefits must be equal to the required risk premium.

If however, the necessary risk premium is not reduced by the same proportion as the variance in the marginal benefits, then the tax system will not be neutral in its effect on capacity decisions. If the required risk premium falls by proportionately more (or less) than the fall in variance of marginal benefits, then capacity will appear more (or less) attractive and more (or less) North Sea activity will be likely. Risk-return analysis may be employed to portray capacity decisions. The effect of taxation is to shift the opportunity locus towards the origin. If the pre-tax choice were point A in Figure 3.5, post-tax this would appear as point A', where A' is on the ray drawn between A and the origin such that AA'/OA is the tax rate. The slope of the opportunity locus at A' is the same as the slope at A. However, the slope of the indifference curve at A' is unlikely to be the same as that at A. If it is flatter then this will indicate a decreased aversion to risk and hence a desire to take on more risky operations; that is, to expand North Sea capacity. Conversely, if the indifference curve is

Figure 3.5

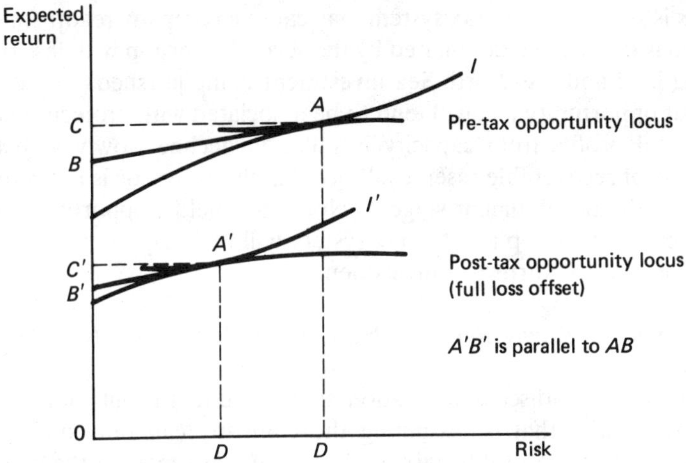

steeper, the required risk premium will have increased proportionately to the size of the expected returns, and greater risk aversity will reduce investment in risky North Sea reserves.

A non-distortionary outcome will occur only when the slope of the indifference curve at *A* is the same as the slope of post-tax indifference curve at *A'*. This, we must emphasise, is unlikely to be the case. Since *A'* is south–west of *A*, the move from *A* to *A'* may partly be a move around one indifference curve. As a south-westerly movement around an indifference curve takes place the slope of the indifference curve tends to flatten and the company will tend to require a smaller risk premium. If utility is held constant, then as exposure to risk is reduced the equivalent marginal certain return needed to compensate for risk will fall. Alternatively, since *A'* lies inside the opportunity locus involving *A*, utility must be reduced. That is, the move will be to a lower indifference curve (from *I* to *I'*). Casual observation suggests that lower income companies will require more compensation for exposure to an additional unit of risk: that is, lower indifference curves will be steeper. The ultimate outcome will depend on which of these two effects dominate; while the former effect will always tend to encourage additional risk-taking, the latter effect is ambiguious and may encourage, although it seems more like to discourage, risk-taking.

In general, the outcome will be related to the shape of the

opportunity locus. If the locus is very flat then as a result of taxation the company is likely to become less risk-averse. Alternatively, if the locus is steep then the tax system may cause a company readjustment which is likely to be dominated by the second assertion with less risks being held and less North Sea investment being pursued.

Flat opportunity loci will tend to be associated with projects where marginal profits from capacity installation decline slowly with the volume of recoverable reserves. Since this characteristic is not prominent in the development stage of North Sea fields, opportunity loci will tend to be steep and the tax system will be likely to operate as a disincentive to North Sea investment.

The Effect of Taxing Revenues but not Tax-Relating Costs

So far we have discussed taxation as if it reduces profits and losses symmetrically: that is, assuming the company can fully offset tax losses against the profits earned elsewhere. This may be the case as we are considering the marginal profit or loss where intra–marginal units will generate positive tax profits. However, the scale of offshore finances and the volatility of the economic climate around North Sea profits both tend to make a company's North Sea assets dominate its portfolio. This implies that the tax profits on non-North Sea operations are relatively small and that intra–marginal units may not be unexposed to making a loss. Hence, the tax system may operate as a tax on profits with no offsetting tax subsidy available on losses.

Without full tax offset, the riskiness of a North Sea venture will not be altered by imposing tax (since we consider risks to be measured by the exposure to possible losses), although the expected return will decline. The effect of taxation will be to lower the slope of the opportunity locus relative to the slope with full tax offset, because for a fixed increase in the expected return greater additional risks must be borne; this is shown in Figure 3.6 where the no-loss-offset opportunity locus (BB'') is both lower and flatter than the full-loss-offset opportunity locus (BB'). It is worth noting that the peak of the locus will be associated with less North Sea capacity.

Flattening the opportunity locus will make companies more reluctant to invest in risky North Sea projects and taxation will tend to act as a deterrent to installing capacity. We should note that while the disincentive compared to fully offsettable tax is unambiguous, there still remains some ambiguity in the ultimate outcome compared to

Figure 3.6

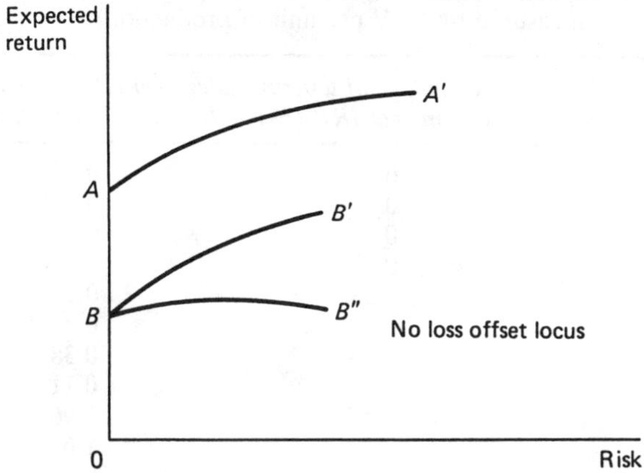

pre-tax; the possibility that a tax system will encourage risk-taking is, however, even more remote.

Evidence

Companies investing on the scale necessary to develop an offshore oil field must be wary of events that could undermine their expected returns. Initial appraisals of field economics must incorporate a level of expected profits to compensate the firm bearing risks. These risk profits will be most crucial on the less profitable developments: the more promising reserves are likely to be worthwhile even under pessimistic forecasts of oil prices and costs. We are therefore crucially concerned that PRT does not impose an additional burden on the marginally commercial projects.

The tapering and safeguard restriction on PRT is designed to put a limit to the tax burden and a floor to possible profitability. By grouping commercial oil fields according to their (pre-tax) profitability, Table 3.5 assesses the usefulness of the tapering clause to each of the established commercial fields. In general the tapering clause is or has been helpful to some but not all of the less profitable fields; however, it also has been an advantage to several projects not needing protection (Claymore, Thistle, Ninian) and cannot guarantee to protect all of the least attractive fields (for example Heather, Tartan). Moreover, of the 24 potentially commercial fields examined,

Table 3.5 The impact of the tapering and safeguard provision in commercial oil fields ranked in order of their pre-tax profitability (as measured by NPV per unit of production)

	Advantage of tapering safeguard (10% discount rate)	
	(a) on real IRR (%)	*(b) on real NPV (£/tonne)*
Piper	0	0
Forties	0	0
Argyll	0	0
Murchison UK	0	0
Claymore	0.58	0.7
Auk	0	0
Dunlin	0.28	0.38
Fulmar	0.15	0.11
Ninian	2.57	2.96
Thistle	1.09	1.6
Beryl	0	0
Brent	2.68	3.31
Fulmar	0	0
Brae	1.64	1.58
Duncan	0	0
Stratfjord UK	0.73	0.79
Cormorant North	2.11	2.74
Cormorant	0	0
Maureen	1.6	2.69
Heather	0	0
Montrose	0	0
Magnus	2.49	2.68
Hutton NW	1.59	2.08
Beatrice	1.96	3.24
Balmoral	0	0
Brae North	0	0
Hutton	2.21	2.95
Alwyn North	2.05	2.44
Highlander	0	0
Clyde	0.18	0.25
Tartan	0.01	0.01

only Bruce and Tern are affected by the tapering and safeguard provision. The restricted time period for use of this provision means it only operates when other allowances, particularly the oil allowances, are already shielding revenues from tax.

Tapering alters the impact of a decline in oil prices on project returns. We have assessed the decreases to project profitability caused by a decline in oil prices, first with all the PRT allowances and then with the tapering and safeguard clause removed. More often than not, the tapering clause makes no difference or only a very small difference to the impact of depressed prices, despite the hopes for this provision. The moderating influence is strongest for the fields of slightly less than average profitability where project returns drop less, as a result of a price decline, than they would without tapering. The decrease to most of the less profitable fields' profits per barrel is reduced because of tapering. However, if events do change for the worse, the decrease in expected profitability for some of the less profitable projects is not guaranteed to be lessened because of tapering, and this clause actually exaggerates the impact for Maureen (the decline, of course, is from higher post-tax unit profits when tapering is included.)

The tapering clause does not limit the impact of PRT effectively for all the least profitable fields even though the effect of falling prices tends to be more severe for the less profitable discoveries than for the more desirable projects. Brae, Buchan, Hutton, Maureen and Cormorant North are likely to be liable to large sums of PRT despite the existence of tapering. Conversely, the PRT liability of some of the better projects is restricted to barely significant amounts. Tapering does little to stop PRT deterring the exploitation of marginal reserves. Tapering at best only partially protects a project's risk profits and PRT is likely to discourage companies from investing in the North Sea, although they might have been willing to invest and bear the risks in the absence of PRT.

Lessons

Different tax systems will offer different incentives for holding incremental units of risk. In the current North Sea system, taxation is unable to distinguish between risk profits and pure profits. As a result, not only is the opportunity locus facing a company shifted towards the origin (the inevitable consequence of any tax, showing that the Treasury shares in the risks as well as taking a part of the expected ration), but the opportunity locus is also flattened (showing that for a given increase in expected returns greater additional proportionate risks must be faced, because the Treasury shares in risks by less than its proportionate take of profits). The flatter the opportunity locus, then, for a given increase in risk-taking, the expected return increases by a smaller amount. Hence companies are discouraged from bearing risks and North Sea developments are deterred. For the opportunity locus not to flatten, taxes as a proportion of profits must decline as capacity increases and as marginally less attractive barrels are added to each field's recoverable reserves. It is because such progressiveness is lacking from the North Sea tax system that the topic of risk profits becomes important to development decisions.

One lesson that has emerged from this section is that the tax system should participate symmetrically in risk profits and risk losses as they affect the company. That is, first, the risk profits/losses should be defined to reflect the company's opportunity cost of capital and second, they should reflect a company's risk-aversity by recognising the more significant consequences of making losses against the relatively smaller gain from an additional unit of profits. For this to occur, the oil taxation system should leave marginal profits untaxed and be progressive with respect to individual field profits. Furthermore, the tax take should be sensitive to prices as otherwise the impact of taxing risk profits will be aggravated.

3.4 TAXING INCENTIVES

Definition

In the offshore oil sector concern that the oil companies will receive windfall profits has arisen whenever oil prices increased. In order to capture at least a part of this apparent windfall, the government has increased the level of taxation. The economic rationale for taxing windfall gains is that the windfall, by definition, does not lead to any change in economic activity. In a North Sea context, though, the windfall nature of additional profits is questionable as company plans will consider various oil price trends. Hence sharp increases in oil prices are unlikely to be totally unforeseen and the possibility of an oil price rise will be relevant to companies' development decisions. Any attempt by government to tax those profits will tend to act as a tax on incentives and thereby to deter development plans.

Successive UK governments have acted as though they were unaware of the reasons why oil prices rise or the role of oil company profits, and history shows a strong connection between increases in the oil price and increases in taxation (see below). Oil prices, however, rise for the fundamental economic reason that there is a shortage of oil worldwide, and will continue to rise so long as oil demand exceeds supply. The rising price then acts as a signal both to reduce demand and to increase supply. By interfering with the incentive to increase supply, the government may jeopardise this desirable supply response.

The difference between progressivity and increasing taxation as prices change should be made clear. Progressivity refers to the tax take increasing as pure profits increase, at a more than proportionate rate, and leaves marginal incentives unaltered. On the other hand, increasing taxation by response to an oil price increase is concerned with structural changes in the tax system which imply a new set of marginal incentives.

Relevance to Development Decisions

In the event of oil demand exceeding supply, rising prices, in an economy directed by market signals, will tend to alleviate the shortage. On the one hand demand will be restrained, and on the other hand previously uncommercial supplies will become viable. In the UK oil industry this implies that as oil prices rise the economy will benefit by developing some oil fields which formerly had been considered too costly, instead of buying more expensive (in terms of resources) imports.

While rising prices indicate the need for more investment, higher profits will provide the resources and higher expected profits will provide the incentives for more investment. In a competitive environment, profits direct companies away from industries where the price-cost margin is low into activities where the price-cost margin is higher. With higher oil prices, oil-related profits are likely to rise relative to non-oil profits and extra resources will be moved into oil – field development. Such movements benefit the economy at large, shifting capital and labour away from low value added uses into more rewarding employments, and increasing the net value of aggregate productive activity.

We do not assert that market incentives yield in any sense an optimal allocation of resources in the UK offshore oil sector, but at least they reallocate scarce resources in the correct direction and alleviate imbalances between demand and supply when pertinent economic factors change. This desirable reallocation presumes market incentives are allowed to direct resources.

The UK government, however, appears to view oil company profit increases as 'windfall' gains without recognising the long-run changes that such profits induce. We contend that this government response is an ill-conceived move based on a myopic view of how the oil industry operates. Companies planning to invest large-scale funds are wary of factors which might undermine their revenues in years ahead. Once the government believes its responsibility is to tax away all or most of the profit increases, the desirable link between higher prices and increased supplies is broken. With fewer resources and reduced incentives, the oil companies will be less eager to develop new North Sea oil fields. Indeed, the impact of the tax changes on company expectations of government policy could be severe.

A government policy of raising extra North Sea tax revenues may help the government to reduce its Public Sector Borrowing Requirement (PSBR) in the immediate future, but such a policy conflicts with longer-run UK oil industry needs. More importantly, if government's myopic view remains unchanged and if government continues to concentrate on short-term objectives to the exclusion of long-run matters, then the benefits of UK oil production to the UK economy may be severely eroded. Perhaps a lack of understanding has led to government myopia, but more probably government understands the issues but either does not appreciate the extent of the damage likely in the 1990s or has chosen to ignore such undesirable long-run consequences of policies with short-term political appeal.

Evidence

Tax Changes as Governments' Response to Oil Price Changes

Amending North Sea taxation is by now the familiar British government response to rising oil prices. Support for these tax changes comes from all political parties and receives favourable comments from the press. This section establishes that government policy is to raise North Sea tax rates or lower allowances as oil prices rise. Seven periods of taxation can be identified – the original system plus six groups of changes – which have been set out in Table A.1 and are explained in Appendix A. We now trace the history of North Sea taxation from its origins in 1975 to demonstrate the link between oil price changes and tax changes, and in particular to demonstrate that this link is a deliberate intention of policy.

When the then Paymaster-General, Mr Edmund Dell, finalised PRT in 1975, he was confident of the impact of the tax:

> By this tax we are trying to achieve a great and irreversible switch of profits from the oil companies to the British Government. However . . . we have to accept that the oil companies must make a reasonable return on their investment. That is all we have done. We have established a reasonable sharing between the British people and the companies.[4]

Considerable concern was expressed that the system should protect company profits if oil prices failed to hold their contemporary value, a concern of which Mr Dell was aware: 'I have given careful thought to the representation made that there should be some automatic protection against a fall in oil prices and a consequential reduction in revenues. We therefore propose that there should be a safeguard provision.'[5] He continued to explain that: 'The clause has been introduced to provide some automatic protection against a fall in the real price of oil, or what amounts to the same thing, movements in domestic prices unfavourable to oil.'[6]

Upwards movements in the price of oil do not seem to have been catered for in PRT automatically. The Paymaster-General, however, made clear the action that would be taken if oil prices rose faster than the commonly held expectations for depressed oil prices (outlined below).

> I am, of course, fully seized of the fact that a change in the price of oil relative to price levels generally could have a profound effect on profitability. I therefore made it clear on Second Reading, and I take this opportunity to reaffirm that the Government will stand ready to review and adjust the incidence of PRT in the event of a sustained and significant change in the price of oil in real terms.[7]

Although the passage of PRT through Parliament generated much debate,[8] the lack of debate on the necessity to amend PRT in the light of movements in the real price of oil reflected a consensus of view. By 1978, the then Chief Secretary to the Treasury, Mr Joel Barnett, considered the oil price to have moved significantly from its 1975 level:

> We have come to the conclusion that there is scope for increasing the share of North Sea profits which accrue to the public. When we fixed the rates and allowances we deliberately adopted a cautious

approach. We set the rate of PRT no higher than 45% and we gave generous allowances and reliefs because of the great uncertainties of the time. Now, however, though many uncertainties remain we are in a position to take stock and it is apparent that companies are obtaining very large profits from the national resources of the nation. We believe that the public share of these profits can and should be increased without endangering the exploitation of the less well-placed fields.[9]

The same conclusion was drawn by the then Conservative opposition spokesman on energy, Mr Tom King: 'We accept that there is clear scope for adjustment of the tax regime.'[10]

Once the Conservatives were in office, the previous administration's proposals for tax changes were legislated in the June 1979 Budget. Talking about the proposals announced in August 1978, the new Chancellor, Sir Geoffrey Howe, said:

> I have judged them now against the background of recent rises in the price of oil. On that basis, the original package of PRT proposals for giving the Government more revenue from the North Sea is now fully justified. Accordingly there will be provision in the Finance Bill to implement it.[11]

Early in 1980 the situation had changed further. The Chancellor was not oblivious to the changes in the oil industry.

> Since my last Budget, World oil prices have increased dramatically.
>
> North Sea oil prices, which follow world prices, have risen by more than half, from about $20.70 to $33.75 a barrel. This substantial change has greatly favoured the oil companies. I therefore propose . . . to increase the rate of petroleum revenue tax. The increases in oil prices have greatly strengthened the industry's cash position.[12]

The Chancellor went on to introduce a higher rate of PRT, which received no opposition from the Labour Party. Mr Callaghan replied: 'There is much to be said for increasing the petroleum revenue tax from 60% to 70%.'[13]

Later in the same year, however, further tax changes were announced by the Chancellor:

> Since March there has been time to assess more fully the implications of the steep rise in the price of oil since 1978. This has been on as large a scale as that in 1973–4. The government have concluded that there is scope for a further increase in oil taxation. Petroleum Revenue Tax – because of the way reliefs are structured – does not provide an adequate means of obtaining further revenue. I shall therefore introduce in the next Finance Bill a supplementary tax to be paid in addition to PRT, to take effect from the beginning of next year [1981].[14]

On behalf of the opposition party, Mr Denis Healey, the Shadow Chancellor, concurred with this view: 'The Opposition welcome the Chancellor's announcement that he will put a tax on the windfall profits of the oil companies.'[15] Hence in March 1981, Supplementary Petroleum Duty (SPD) was announced by the Chancellor:

> Increases since 1978 in the real price of oil have brought substantial benefits to the oil companies, which face a very different prospect from that when the present tax regime was introduced. Such has been the rise in the oil price in recent years that I believe the Exchequer should properly look to this area for additional revenue beyond what will accrue from existing taxes.[16]

Moreover a proviso announced in 1975, to limit the use of PRT allowances, was also enacted in the March 1981 Budget. This proviso was made by Mr Dell:

> One can hope that at any rate by the early 1980's the present uncertainties surrounding the price of oil will have been resolved and that the relationship between oil prices and other prices will be seen on a firmer basis . . . It would be for the Government of the day, in circumstances, one hopes, of greater certainty about relative prices than now exist, to decide whether they should permit a field of this kind [(profitable)] to benefit from the safeguard provision after it had yielded substantial profits on the investment earlier on.[17]

Not only was the safeguard provision limited to a period half as long as the period to PRT payback, but from the start of 1981 uplift was no longer permitted after PRT payback. Oil prices, however, did not maintain the growth witnessed between 1979 and 1981. At the start of 1982, current oil prices were depressed and the forecasts for rapid real oil price growth throughout the 1980s and 1990s were revised downwards.

The Chancellor reviewed these changes in his March 1982 Budget speech.

> Last year, in the light of the massive increase in oil prices which had occurred in earlier years, we changed the structure of North Sea tax, to make it more responsive to changes in price. At the same time tax revenue from the North Sea was brought forward, with an increase in the total level of taxation.[18]

The fall in oil prices did not raise official concern:

> The current fall in oil prices reduces the revenue that the Exchequer receives. I recognise that it reduces the revenues of oil companies as well – but it also reduces the tax they have to pay. Detailed study has convinced me that . . . the total tax burden is not such as to discourage exploration or development. Nor is it so high as to deprive the oil industry of a reasonable, and often attractive, yield. In these circumstances, I cannot reduce the overall tax burden to the extent that the industry would have wished. But I do agree with it on the need for some change of structure.[19]

Without any serious debate the Chancellor's measures were legislated. Advance Petroleum Revenue Tax (APRT) replaced SPD and the rate of PRT was increased to make up the bulk of the shortfall in tax revenues which might otherwise have occurred.

The history of North Sea taxation has seen numerous substantive changes since PRT was first formalised. Continuing his Budget speech, Sir Geoffrey referred to these changes:

> I am aware of the concern felt by the industry about the number of changes in the regime there have been in recent years. For this reason, my hope is that the new tax structure that I have proposed will provide a more secure and stable regime for the future, permitting development to go ahead uninhibited by major fiscal uncertainties.[20]

This hope was dented some three months later when, in June 1982, adjustments to the new system of taxation (which had then not even been in operation) were announced: 'I believe there are adequate returns on North Sea projects at the current broad levels of taxation. do, however, accept that the operation of my proposals, particularly as they affect some more marginal fields, could be improved by modification.'[21] In his press statement issued to explain the modifications, the Chancellor confirmed the connection between tax changes and oil prices: 'It has been an opportunity to take further stock of the position in the light of the latest prospects for oil prices.'[22]

The 1983 Budget in March occurred in the wake of OPEC's London meeting. Amidst predictions of a dramatic oil price fall OPEC finally agreed upon a $5 decrease in the OPEC reference price from $34 per barrel to $29 per barrel. In his budget statement the Chancellor explained the large-scale tax changes:

> Tax is not the only factor sustaining North Sea potential. Steps taken by the industry to cut costs and the future level of oil prices will be at least as important. But the tax structure must adapt as well.
>
> I am therefore proposing a substantially more favourable regime to assist the companies as they move on to develop new fields.[23]

Substantial changes were made to the oil tax regime in response to falling oil prices and falling oil price expectations.

Mr Lawson, in his first Budget as Chancellor of the Exchequer, announced in March 1984 that oil tax changes were being considered. Again, oil prices had been falling in real terms and the expectation of more price reductions remained: 'The Government are already committed to a study of the economics of investment in incremental development in existing fields. This is of increasing importance . . . I therefore propose to review this area with the industry, and to legislate as appropriate next year to improve the position.'[24]

Taxing Incentives in the North Sea Oil Industry

The following sections consider the way the oil industry operates, the government view of how the industry operates, and divergences between the two.

Although the government has consistently stated a desire to collect a 'fair' share of North Sea profits, its notion of 'fair' seems to change to suit changing political interests. The precedent is clear and now companies probably expect further tax changes if oil prices continue to rise. Once the private sector begins to expect further government interference, one of two possibilities is likely.

Outcome of Oil Prices Rise

Even if oil prices continue to rise, oil investment may be deterred by government who change the tax system in response to higher oil prices. One desirable consequence of rising oil prices is that an impetus will be given to oil investment and fields which had previously been non-commercial will be developed. This, however, is based on two prerequisites: that the capital funds exist and that the incentives are sufficiently improved. In the North Sea, however, it may be that the first of these prerequisites is broken because even if oil prices do rise there will be negligible additional funds for investment in new fields since the level of taxation is likely to be increased in line with oil prices. Be examining aggregate cash flows of established commercial fields it is possible to estimate the scale of equity sources which might be available for North Sea investment in the future.

It would be expected that as a result of an oil price increase the North Sea industry would be, *ceteris paribus*, more profitable and there would be ample finance available to expand and develop new oil fields. However, this consequence of higher oil prices ignores oil tax changes. Above we demonstrated that an increase in taxation tends to be associated with a price increase. Incorporating tax changes associated with higher oil prices into calculations of aggregate net cash flows, it is apparent that the effect of the tax changes is to moderate the increase to industry cash flows resulting from increased oil prices.

Whereas the preference for higher oil prices is a preference that one would expect the North Sea industry to demonstrate, this preference may not longer hold if oil price changes necessarily induce tax changes. It may be the case that the industry might actually prefer lower oil prices (and less harsh taxation) to higher oil prices (with harsh taxation). This reversal of preferences because of tax changes is possible but not definite. What is definite is that oil price increases tend to bring forth tax changes which nullify the advantages of higher oil prices and the increased attractiveness of North Sea investment will be attenuated, the expansion of the North Sea industry will be moderated and companies' keenness to develop new fields will be lessened.

Early tax changes in 1978 and 1980 had a small impact relative to later changes when the effect has often been substantial and has sometimes even reversed the direction of the impact on cash flows. Perhaps the best example of this was the 1981 tax changes (when SPD was introduced, uplift was limited to a period up to PRT payback and tapering and safeguard concessions were reduced). These changes more than offset the impact of higher oil prices on development activity (this is quantified in Chapter 4). On the other hand, the 1983 tax changes (at a time of depressed oil prices) had the effect of reducing the overall level of taxation but not by reducing significantly early tax revenues. The 1983 experience shows a ratchet effect on aggregate industry funds in operation. When oil prices rise the tax system becomes harsher; but when oil prices fall, tax revenues, and in particular early tax revenues, are not significantly reduced. The moral of this section is clear; if oil prices rise, company cash flows will not necessarily improve, funds available for North Sea investment may not be forthcoming from internal sources and the expansionary impact on UK oil supplies might not materialise.

A field-by-field analysis for a range of possible new discoveries emphasises the possibility that the second prerequisite for the desirable supply response – sufficient incentives for new project developments – may also be broken by a policy of increasing the level of taxation in response to higher oil prices. So long as there are no tax changes, higher oil price expectations will lead to a rise in the real NPVs per tonne on new oil fields. After considering tax changes, however, the impact of higher oil prices on unit profits of new discoveries is severely diminished.

If oil prices do rise rapidly, then smaller increases in field-by-field profitability will be more likely than would occur in the absence of a policy connecting the level of taxation to oil prices. Given that expected return would be reduced, exploration enthusiasm would be dampened and some of the additional supplies that might be forthcoming in the event of rapid oil price growth would not be produced. For instance, it may be the case that with a relatively high oil price path the NPV per tonne of 'typical' new fields may be reduced by between one and two percentage points as a result of the level of taxation increasing with oil prices. Furthermore, the ratchet effect associated with oil prices and tax changes is an important consideration here.

Outcome if Oil Prices Fail to Rise

The government's policy of raising taxes in response to higher oil prices is likely to have an undesirable consequence even if oil prices do not rise.

Once history has established that governments will increase the tax take whenever oil prices rise then there will be an impact in the future even if prices do not continue to rise and there are no further tax changes. We have argued that oil price rises may no longer bring forth additional UK oil supplies because aggregate post-tax cash flows and after-tax field rates of return may not improve with higher oil prices. Hence it may be the case that in the event of no further growth in oil prices, North Sea production may not be as great as it would have been if there had been no history of tax changes. After-tax cash flows and unit profits will be lower than they might have been if the tax system had not been altered since 1975, which suggests that North Sea oil activity will run at a lower level than would have otherwise occurred. Furthermore, a government policy of linking the level of taxation to oil prices will create a future with rather one-sided risks for oil companies (see section 3.3 above).

Before an oil company undertakes an exploration or development project in the North Sea it goes through a complicated process, evaluating the factors which may affect the profitability of its investment. The company might use its contemporary forecasts of inflation, exchange rates and real costs under a variety of oil price scenarios in order to quantify the expected profitability and riskiness of any discovery. Incorporating the link between tax policy and oil prices might alter this evaluation process dramatically.

The chances that oil prices may be lower than some central estimate, and that losses for the companies involved in these discoveries might arise, are unlikely to be significantly affected by a less severe tax regime. However, the chance of earning substantial profits, with higher than forecast oil prices, is affected markedly with possible unit profits and rates of return unlikely to rise substantially due to tax changes. The commercial attractiveness of all new projects is likely to decline, and the likelihood of each project being pursued is lessened, because of the dynamics of taxation. For example, by looking at the improvement to NPVs per tonne due to changing oil price expectations with and without tax changes on assumed new fields, we may estimate that tax changes reduce expected profitability from the level it otherwise might have been between 0.5 per cent and 1.5 per cent.

Oil companies in the British Sector of the North Sea face a one-sided future; a future where there is a chance of possible oil price falls and associated losses but there no longer seems to be much chance to gain from oil price rises. Even if oil prices do not rise rapidly, the tax changes make it unlikely that companies will explore as intensively or develop as many fields as they might otherwise have done.

Lessons

A government policy of continually meddling with oil taxation adds a disincentive to long-term company planning which will reduce North Sea investment and consequently North Sea oil supplies. Government is either unaware of the long-run damage caused by interfering with the profit incentive or is so dominated by short-term considerations that it has little concern for long-run adverse effects.

A fundamental lesson of this analysis is that the government must appreciate the importance of oil company profits in times of oil price changes. The volatile nature of oil prices makes it unlikely that price increases are entirely unforeseen and entirely unanticipated, and thus profits must be allowed to rise if the positive impact on development decisions is to occur. Governments must understand that it is not possible to alter the distribution of profits by interfering in the profit mechanism without distorting the decision-making process, which could have dire consequences.

It is crucial for government interference to be minimised. This objective would be helped if the tax system automatically increased the tax take in response to higher oil prices (that is, as fields' profits increase). This will lessen the need for tax changes to be introduced on the basis of the short-term consideration of the PSBR.

If, and only if, the political pressures are such that adjustments to the tax system still have to be made in response to oil price increases, then the structure of taxation should not be altered. Providing the tax system protects the normal returns to capital and risk premia, then limiting tax changes to tax rates, although undesirable, will have less detrimental consequences than altering the tax structure since this will not compound other faults in the tax system. If these conditions are not satisfied then this second-best alternative should be avoided.

3.5 INSTABILITY

Definition

Tax instability in the offshore oil sector refers to unpredictable government imposed tax changes. No other sector in the UK economy has been subject to such instability. In the UK oil sector, however, there is uncertainty both as to when tax changes will occur and as to what tax changes will occur. Tax instability is an additional source of risk in the offshore industry which imposes additional costs on oil companies.

We first establish why tax instability is important in the oil extraction industry and then outline how instability affects North Sea oil economics. Here we do not appraise the suitability of the tax structure, nor do we assess the fairness of the split of North Sea related benefits between companies and government. Our concern is with the damage to oil industry prospects and confidence from oil tax changes. Finally, we note the lessons to be learnt from the oil industry's experience of tax instability.

Relevance to development decisions

Our interest in this section is with tax changes *per se*. Although we have argued in previous sections of this chapter that UK oil profits taxation does not match the peculiarities of the North Sea extraction industry, so that taxation needs to be restructured, the tax changes which have been implemented were not designed specifically to improve the structure but were merely a response to changed government needs.

It is important to make clear the distinction between taxing incentives (in the previous section) and instability. Section 3.4 established the connection between tax changes and oil prices, whereby increases in tax rates or reductions in allowances tend to occur as oil prices increase. This link should be seen as a predictable government response following from the (misplaced) intention to tax increased profits following oil price increases. Such an intention is misplaced because it results in governments taxing incentives. Section 3.4 demonstrated that changing crude oil prices have been, and presumably will continue to be, a sufficient condition to generate tax changes. As such tax changes along these lines are relatively predictable, they cannot be characterised as adding uncertainty to the economics of offshore oil. However, while history shows us that oil price changes have always preceeded tax changes, so that a significant change in crude oil prices seems to be a sufficient condition for tax changes, a crude oil price change is not a necessary condition and is not the only event that will generate oil tax changes. Indeed, whenever the capacity of the UK oil industry to bear taxes alters in the eyes of the electorate, the government is likely to respond to such changed perceptions by adjusting oil taxation, especially in the months close to an election, in its attempt to maximise votes. Similarly, the process of political lobbying is likely to see oil taxation traded for other policies in ways which are unpredictable.

Instability, then, refers to *ad hoc* changes in the level or structure of taxation that are not easily predicted. The unpredictable nature of these tax changes creates a further level of uncertainty which increases the riskiness of North Sea investment.

The evidence of the following sections suggests the tax changes will not just change the split of North Sea benefits between oil company profits and government tax revenues, but will reduce the benefits to both recipients. The two principal sources of tax instability – first, political and bureaucratic pressures, and second, the failure to tax incentives – are considered next. We find the former source creates uncertainty over when tax changes will occur and the latter creates uncertainty over both when and what changes will occur.

Evidence

Political and Bureaucratic Pressures

In order to maintain political power, governments like to be seen to be in control of the oil sector. In their desire to maximise votes, governments will attempt to adopt policies that are popular with the electorate on all important issues. For a sector as vital as North Sea oil, this results in governments forming policies which are perceived to control foreign-based multi-national oil companies, to protect a 'valuable national asset' or to guarantee secure oil supplies for the domestic economy. Generally, this pressure is most intense at election times or when the oil sector is at the forefront of the public's attention: for instance, an internationl oil supply crisis receives considerable media attention and politicians introduce or modify policies in an attempt to be seen to be in control.

In addition, government bureaucrats often favour a changing, complex policy in order to reinforce their status and prestige. The more complex a policy and the more times it is changed then the more their expertise becomes exclusive and is demand. Hence by advocating and promoting change to an already complicated tax structure, the bureaucrat's job security and status are protected. As a result of political and bureaucratic pressures on government, seemingly random tax changes occur which are not, at least by oil companies, predictable.

Taxing Incentives and Instability

Changed Expectations

While it is important not to describe government intentions to tax incentives as increasing instability (because the intentions are predictable), the increases in tax rates or reductions in allowances following oil price changes have an element of instability in them. The government's intention is to increase taxation in line with oil industry profits, which it views as synonymous with oil prices. Section 3.4 argued that in principle this link may have detrimental consequences. In practice this link may also create instability for two reasons. First, oil prices are not the only major factor contributing to oil industry profits. Instead oil price moves should be analysed alongside the paths of exchange rates and inflation. As this has not been the case, additional uncertainty over when tax changes will occur has been created. Second, the cumbersome North Sea tax structure has meant that any tax changes have had vastly different impacts across the range of fields. This creates additionl uncertainty for any company who will not know if their fields are to bear the brunt of higher taxation or if their fields will be minimally affected: that is, additional uncertainty about what tax changes will occur is created.

Consider first the instability concerning the timing of tax changes which occurs because of differences between government perceptions of the oil industry's profits and the industry's actual profits. These profits are not only influenced strongly by dollar oil prices, but are also heavily influenced by other key economic factors such as the sterling-dollar exchange rate and the rate of cost inflation. Moreover, the current levels of these variables are most significant not *per se*, but for the indication they give to the future course of each variable. Evidence presented below suggests that the government fails to consider the other key variables and bases its policy proposals for tax changes on dollar oil prices alone. Even if the industry recognises the government will always seek to tax 'windfall' profits, since industry profits are determined by factors other than oil prices the government intention to tax incentives will embody a degree of uncertainty.

By concentrating on the relationship between the industry's performance and the tax changes we recreate the attractiveness of investment in the industry in each period when North Sea taxes were changed between 1975 and 1984. As the oil industry is characterised by some long lead times, at any given time its attractiveness will be based not on observed data but on expectations about the factors affecting its future. The expectations pertaining at any given time must then be used to recreate the industry's attractiveness in past times, in the absence of information and events that have happened since. The expectations for three variables apart from taxation are particularly relevant: the dollar oil price, the sterling-dollar exchange rate and the rate of cost inflation. Contemporary views on these three variables at seven different dates are shown in Table 3.6.

We have recreated the expectations of past periods to explore how expected industry profits were changing and how significant the tax changes have been relative to the move in profits. Although it would be easy to find forecasts that disagreed with the views we have taken, since our analysis intends to look for order-of-magnitude impacts rather than precise quantification, and since we have made our assumptions clear, we do not justify the expectations further.

Changed Aggregate Profitability

Section 3.4 of this chapter, on taxing incentives, suggested that taxing profit increases would be damaging in principle. However, because tax changes have been made in response to nominal dollar oil prices rather than sterling oil profits, the government's attempts to tax profits increases have tended to fail and instead have increased taxes when the improvement to incentives was questionable. The government does not seem to have fully appreciated the range of economic factors affecting the industry or has chosen to subjugate them to more immediate political considerations.

In either case, the government acts as if it has a myopic view of the factors affecting oil profits and investment. Whereas the oil industry's fortunes are more related to the change in real profits between different times, the government acts on different criteria, focusing on the change in the nominal dollar oil price alone. This creates two differences in perceptions between the industry and the government. First, the unit of measurement adopted by the government may be nominal rather than real sterling. This emphasises not only the government's relatively short time horizon but also its appreciation of public perceptions which are largely based on nominal prices. Second, the government tends to analyse the effect of oil price changes while ignoring the effect of any changes in other economic conditions. For example, if the prospects for inflation deteriorate or if

Table 3.6 Changed expectations

		Central expectation						
	Actual	A 1975	B 1978	C 1980	D 1981	E 1982	F 1983	G 1984
(a) North Sea oil prices ($/barrel in current prices or nominal growth rates)								
1975	12.5	12.5						
1976	13.2	12.5						
1977	13.9	12.5						
1978	13.7	12.5	13.9					
1979	19.1	12.5	14.6					
1980	34.5	12.5	15.3	24.0				
1981	36.4	12.5	16.1	25.2	36.4			
1982		12.5	16.9	26.5	38.9	32.0		
1983		12.5	17.7	17.8	41.7	33.6	30.0	
1984		12.5	18.6	29.2	44.6	35.3	31.95	30.0
1985		12.5	19.6	31.2	48.1	37.8	34.03	31.50
1986 onwards		0%	7%	7%	8%	7%	6%	5%
(b) Exchange rate ($/£ current values)								
1975	2.22	2.22						
1976	1.80	2.10						
1977	1.75	2.10						
1978	1.92	2.10	2.00					
1979	2.12	2.10	2.00					
1980	2.33	2.10	2.00	2.20				
1981	2.02	2.10	2.00	2.20	2.20			
1982		2.10	2.00	2.20	2.00	1.85		
1983							1.55	
1984								1.35
1985 onwards								1.45
(c) UK wholesale price inflation (% per annum)								
1975	22.3	15.0						
1976	17.3	12.0						
1977	19.8	10.0						
1978	9.1	8.0	16.0					
1979	12.2	6.0	10.0					
1980	16.3	6.0	8.0	11.0				
1981	10.6	6.0	7.0	8.0	15.0			
1982		6.0	6.0	7.0	13.0	10.0		
1983		6.0	6.0	6.0	11.0	9.0	7.0	
1984		6.0	6.0	6.0	9.0	11.0	7.0	
1985		6.0	6.0	6.0	9.0	9.0	6.5	
1986 onwards		6.0	6.0	6.0	9.0	8.0	6.0	

expectations for the sterling dollar exchange rate move upwards, then *ceteris paribus* the UK oil industry's real cash flows would be reduced.

However, by ignoring changed exchange rate and inflation expectations the government perceives the impact of tax changes as not too burdensome, especially given the improvement to oil revenues following favourable movements in oil prices. The industry, on the other hand, is not only concerned with oil price changes but also with changes to other key economic variables, such as the sterling dollar exchange rate and the rate of inflation. Hence tax changes have often appeared onerous to the oil industry. Indeed the impact of the tax changes tends to be substantial not only upon expected industry real cash flows but, more importantly, relative to the improved cash flows resulting from changed economic variables (the combined effects of changes to exchange rates and inflation).

Historical evidence shows that tax changes have not been linked to industry profitability but to short-term nominal oil price changes and without reference to events which, like rising cost inflation, undermine company profits. The result is that UK oil producers are left with reduced funds at a time when oil prices are rising and, in the absence of tax changes, the market response would be an expansion in this extraction industry. The implication therefore is that either the government misunderstands the determinants of oil industry investment or it is concerned only with short-term revenue. In either case its tax policy seems likely to lessen North Sea production and thereby to harm the UK economy.

An example of the impact of the government's perception of oil industry cash flows is provided in Figures 3.7a and 3.7b. Over the life of PRT, from 1975 to 1982, the change in oil prices alone would have improved nominal cash flows by a steadily increasing amount to the end of the century (the dotted line in Figure 3.7a). This assumes that oil price expectations move from case A to Case E in Table 3.6. The combined impact of oil prices and tax changes between 1975 and 1982 would have improved cash flows by less. The distance between the dotted line and the solid line in Figure 3.7a suggests the impact of the tax changes is noticeable but not too burdensome relative to the improvement caused by more favourable oil prices. This is the view apparently held by the government.

Figure 3.7b represents the view apparently held by the industry. Not only is the industry concerned with oil price changes but it is also concerned with other changing economic variables. Of course, the

Figure 3.7a The difference to aggregate cash flows between case A and case E (all fields)

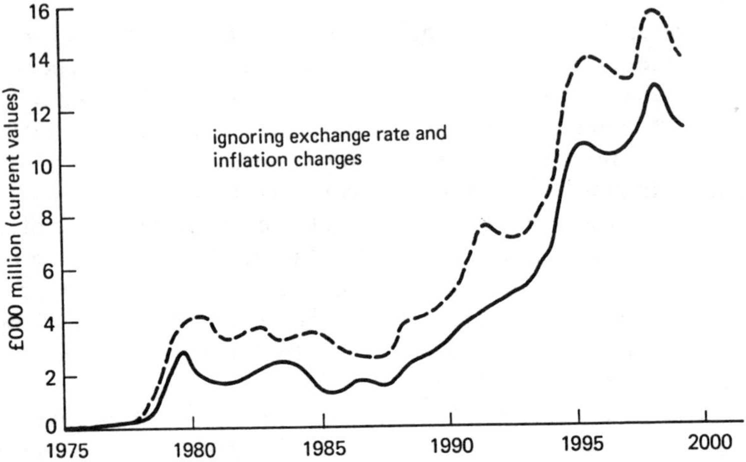

Figure 3.7b

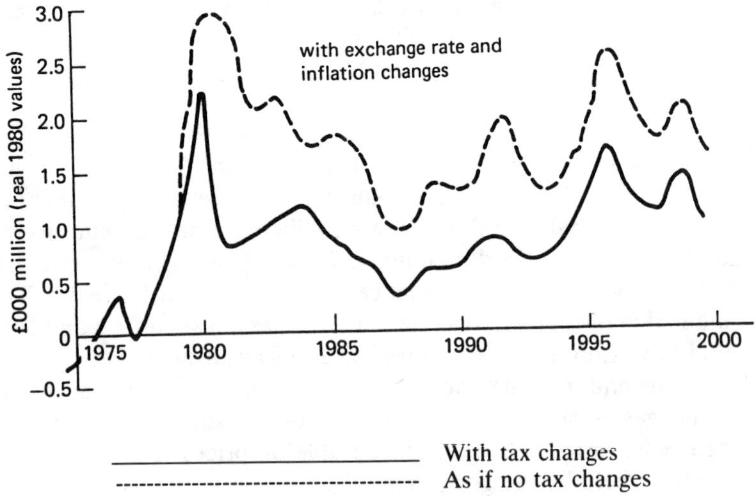

absolute value of the change to industry annual cash flows is much smaller, reflecting the industry's measurement in real terms. The impact of the tax changes relative to the improved cash flows is substantial. In the early 1980s the tax changes account for about two-thirds of the difference in profits, while in the late 1980s and 1990s the tax changes account for about half of the difference in profits. Similar comparisons can be made between most of the periods in Table 3.6.

Changed Field-by-Field Tax Burden

Another way in which tax instability may affect development decisions is through the field-by-field burdens imposed on the industry by changes to the tax system. Apparently the government, tempted by the enhanced returns forecast for the more rewarding oil fields, considers it justifiable to alter the tax parameters because the tax allowances are believed to offer sufficient protection for the less rewarding discoveries. However, sections 3.2 and 3.3 of this chapter showed that the allowances do not protect the returns on fields most needing protection, and section 3.1 showed that the North Sea tax system burdens the less profitable finds while giving relatively favourable tax treatment to the richer oil fields. Whenever tax changes are invoked there is uncertainty over what tax changes will occur and therefore additional uncertainty as to which fields will be affected.

Furthermore, we have found that as the tax level creeps upwards over time the field-by-field distribution of the tax burden has tended to worsen. The more dominant theme since 1975, however, has been that as nominal oil prices rose the tax system has become more onerous and the field-by-field impact of the various changes to tax rates and concessions has tended to be arbitrary. By examining the succession of tax changes since 1975 it is possible to isolate the field-by-field impact of each round of tax changes. For each period we have considered two features: first, the change to the inter-field burden of taxation occurring up to that period, to assess if the structure of taxation has improved or deteriorated, and second, the change to the inter-field burden of taxation relative to the change in field profitability in that period, to assess if any additional burden of taxation has been levied on the fields whose profits improve most.

A comparison of the tax systems in 1975 and 1984 and the surrounding economic climate suggests that the 1984 system for new developments should be more progressive than the system operating in 1975. This improvement could be expected because relatively greater amounts will be collected in cost-related (and therefore profit-related) taxes such as PRT and CT, and less will be paid in royalites which are not related to costs. There are two reasons for this favourable outcome. First, because oil prices have risen to such an extent that field revenues have increased relative to field costs, even though costs have risen. Second, because the rate of PRT net of allowances has risen while the rate of royalty is unchanged for existing fields and less has been eliminated elsewhere, the cost-related elements of taxation have become more significant.

This analysis alone, however, would be too one-sided. Earlier sections explained that PRT operates in almost as regressive a fashion as royalties and, since the changes in first-year CT allowances, CT has become less cost-related. In fact our evidence suggests that on balance progressivity in the tax system does seem to have improved between 1975 and 1984. However, the tax changes since 1975, when analysed in isolation from the changed oil prices, inflation and exchange rates, tend to worsen the inter-field burden of taxation, so that any improvement in tax progressivity occurs not because of the tax changes but despite them.

Associated with the improvement in progressivity, the tax changes ought to have made the least profitable (according to pre-tax unit profits) fields better off relative to the more profitable fields, thereby protecting the normal return to capital and risk premia. Hence lower absolute taxation is warranted for Buchan, Heather and the two Cormorant fields, and perhaps Maureen and Magnus, while Piper, Forties, Auk, Dunlin, Claymore, Murchison UK, Beryl, Fulmer, Argyll and Hutton NW should be the fields to bear the brunt of any additional taxation. However, the impact of the tax changes is as severe on the fields where a reduction in the impact of taxation may have been hoped for as it is on some of the fields where increases in the tax burden could have been made without impairing unit profits too significantly. All of the least profitable fields have faced a greater tax burden, and generally their tax burdens increase more than proportionally than compared with those on the more profitable fields.

A period-by-period analysis of the tax changes and the changed economic climate reveals that after the 1978 tax changes there was a significant change in field-by-field profitability. As the tax burden increased most for the least attractive fields the relative burden became more regressive. However, there is a noticeable similarity between the relative tax burdens with the 1978 changes and without them. This suggests that changes in the economic conditions are responsible for the bulk of the change in relative tax burdens. Similarly, the change in the inter-field tax burden between 1978 and 1980 seems to have been largely due to higher oil prices rather than the tax changes. Subsequent changes in 1981 and 1982 again stress the overwhelming importance of changing economic variables on field economics relative to the rather arbitrary impact of the tax changes. The 1982 tax changes seemed to have slightly improved relative tax burdens and the 1983 changes had an encouraging impact on new field developments: in particular, the abolition of royalties made the new oil fields' tax burden less regressive.

We are forced to conclude that the tax changes, especially those announced in March 1981, aggravated rather than alleviated the inter-field burden of taxation. Furthermore, the evidence suggests that any improvement in tax progressivity since 1975 has arisen despite the tax changes and not because of them.

A second-best approach might accept this failure but ask whether other criteria were satisfied. Specifically, the concern may centre on whether the additional impact of taxation is borne by the fields whose profits are improved most or by the fields whose profits are improved least. Here we are interested in the change to remaining profits per tonne left in the reservoir rather than the lifetime unit profits. Section 3.4 emphasised that taxing incentives can affect project decisions; raising additional taxation when profits increase negates the incentives to develop formerly sub-marginal fields both directly and via the creation of one-sided risks. If the additional taxation is raised from the fields whose profits barely rise, then the disincentive effect will be compounded. Conversely, if the fields whose profits increase because of changed economic conditions are also those that are affected relatively more by tax changes then the disincentive effect will not be aggravated.

There is a noticeable lack of correlation between the impact of changed economic conditions and the impact of changed taxation since 1975. Some fields with the smallest improvements find that the impact of the tax changes is at least as severe as on the fields whose profitability increases a lot. The tax changes seem unable to differentiate according to changes in profitability and consequently the disincentives will be exaggerated.

The lack of correlation between the impact of economic conditions and the impact of tax changes is a common feature resulting from all the tax changes between 1975 and 1984. Between the initial 1975 system and the tax system of 1978 the impact of the tax changes is generally smaller for the fields whose unit profits have increased least or fallen most. This effect, however, is far from consistent. More disturbingly, the tax changes exaggerate the declines in unit profits of all the fields whose unit profits fall. This exaggeration is repeated as a result of the 1981 tax changes, albeit for fewer fields. Both the 1978–80 tax changes and the 1980–1 changes show negative correlation between profit changes in the absence of tax changes and profit changes as a result of tax changes, with the fields benefiting least consistently bearing the brunt of the impact of tax changes. Conversely, when in 1982 tax changes were made after economic conditions had reduced North Sea profitability, the benefit of abolishing SPD was enjoyed more by the fields whose profitability fell least. The disincentives to develop fields, caused by additional uncertainty over which fields will be affected by tax changes, will be great.

The discussion of the field-by-field impacts of tax changes since 1975 emphasises a further problem associated with the instability of the tax system; that is, PRT cannot readily be improved. The arguments presented earlier suggest that adjusting the PRT allowances in order to enhance marginal prospects would probably give a greater boost to the fields least needing help. Similarly, changes in the PRT rate to boost Treasury income from the North Sea are unlikely to have any marked effects.

The indiscriminate and arbitrary nature of changes in the tax burden is most unfortunate. Fields that have benefited little in recent times and can least afford to pay more tax are penalised just as heavily as the more prosperous fields which have benefited greatly. While raising extra taxes from profitable fields reduces the funds for North Sea investment and while the tax changes dissuade North Sea investors from developing new fields by threatening future profits, fields such as Piper, Forties and Claymore may be able to afford higher taxes and still be commercially attractive projects. Unfortunately, such rewarding discoveries were found several years ago and the oil industry may well be right in its belief that future finds will be mainly small size deposits in deeper and remoter waters.

Lessons

Section 3.4 outlined the dangers of attempting to tax profit increases that follow oil price increases since, in principle, this may negate the incentives to install additional capacity. This section has suggested there are additional dangers of altering North Sea taxation in practice; first, because oil prices are not the only factors that play on oil company profits and second, because the cumbersome North Sea tax structure means tax increases tend to fall on the fields which can least afford to shoulder additional tax burdens.

Tax stability would be most desirable. Apart from the additional uncertainty created by governments who seem unaware of the dangers of altering North Sea taxation whenever it is politically expedient – which in itself would be sufficient to deter development decisions – arbitrarily altering allowances and concessions within the oil tax package has a detrimental impact on oil–field economics; progressivity is impaired, the return on capital is depressed and risk profits are penalised.

If, and only if, the tax structure can discriminate between rich and poor fields, then the government should consider raising extra North Sea revenues from the existing tax system. As the burden of taxation is badly misplaced, and as the changed tax rules serve only to accentuate the unfair tax distribution, tax changes seem ill advised. An indiscriminate additional tax burden is not only unfair but also dangerous to the economy. For if the profits from the more complex, more risky and more costly fields are taxed away, companies will have no incentive to develop these reserves, North Sea investment will be lower than it would otherwise be, and potential North Sea output will have to be replaced by relatively expensive imports. The government seems clearly to be raising too much of its extra revenue from the fields which give the companies barely sufficient commercial incentive, and as a result the UK economy will suffer.

Such an *ad hoc* approach has to date back to a two-tier structure of North Sea taxation with a relatively lenient tax system for new fields, to encourage their development, while older fields are subject to a much higher average tax rate in order that government revenues are maintained. Our reasoning, however, suggests that when these new fields commence production the government may tighten the fiscal regime.

Once the new fields produce oil, capital expenditure will be largely sunk, putting oil companies in a relatively weak bargaining position as compared with the government. Furthermore, it is likely that by the 1990s both tax revenues and oil production will be in decline and the government may attempt to moderate both decline rates by the various means at its disposal, including changes to the tax system to preserve its tax revenues. For instance, this may lead to a third category of oil taxation; a new lenient tax system for fields then (in the 1990s) about to be developed, a harsh system for fields commencing production (our 'new' fields) and another lenient system for mature, 'old' fields nearing exhaustion in order to lengthen their production tail. To avoid this undesirable outcome a tax system needs to be adopted which would operate in tandem with political and bureaucratic pressures. Perhaps if a system could be devised that would generate substantial additional North Sea recoverable reserves then politicians and bureaucrats may see their interest served by creating a stable tax environment in order to increase their kudos and influence by stepping up the scale of offshore operations.

If, however, the principal pressures on the government, stemming from the need to be seen to be in control of the oil sector, are so strong that oil taxes must frequently be altered, then a second-best approach should be adopted. If this is the case, then to satisfy government's need to be seen to be in control, perhaps changes to oil taxation can be restricted to changes to tax rates alone, rather than changes to tax allowances. This, though, cannot be achieved within the present system because changing tax rates makes no material difference to the level of taxation. However, if the tax system is highly complex, changes in tax rates may still create uncertainty. Hence a simple system whose tax rates have a meaningful impact on the level of taxation would help tax stability by reducing the uncertainty over what changes will occur. It would, of course, be better if no changes were to occur, thereby also eliminating uncertainty over when changes will be made.[25]

The only remaining rationale for tax changes is if the oil tax system is to be reconstructed in the light of the problems encountered in, and the peculiarities of, offshore oil projects, an approach which should have been adopted in 1975. Greater emphasis would be placed on the plight of the barely commercial discoveries while the better projects would bear more of the burden of taxation. This reconstruction would need to be major, as piecemeal thinkerings would be unlikely to provide any rectification of the problems outlined in sections 3.1–3.4.

4 The Impact of Taxation on Recoverable Reserves

4.1 MODELLING THE SUPPLY OF NORTH SEA OIL RESERVES

In order to quantify the impact of oil taxation on development decisions a behavioural model of the oil extraction industry has been constructed. This model is detailed in Appendix B. It is an econometric representation of the relationship between recoverable reserves in the North Sea and long-run oil prices, for any specified tax system, and can be used to estimate various supply curves for UKCS oil reserves based on different tax systems. Thus, by incorporating different oil tax systems into our model, the impact of taxation on UKCS recoverable reserves can be quantified.

In short, our model uses calculations of the oil price at which each field just becomes worthwhile to develop in order to build a supply schedule as a function of oil prices and to estimate the sensitivity of recoverable reserves to oil prices. Hence, by altering the price at which fields are just worthwhile, the tax system will alter the relationship between recoverable reserves and oil prices, in effect shifting the supply curve. For example, if a tax system becomes more onerous so that marginal profitability declines, then the price at which a field is just viable will rise and fewer reserves will be commercially recoverable at any given oil price (shifting the supply curve as plotted in Figure 4.1 to the right).

The magnitudes of the shifts caused by the various tax provisions are reported in subsequent sections of this chapter. In particular, we consider the shift in the supply curve resulting from the imposition of the current tax system as against a no-tax situation. In general, we believe that the nature of the impact on the long-run supply curve is widely recognised but we doubt if the dramatic scale of the impact is appreciated.

The Nature of our Behavioural Model

Our behavioural model is based on marginal decisions and the impact of taxation described below relates to the impact of taxation on

Figure 4.1

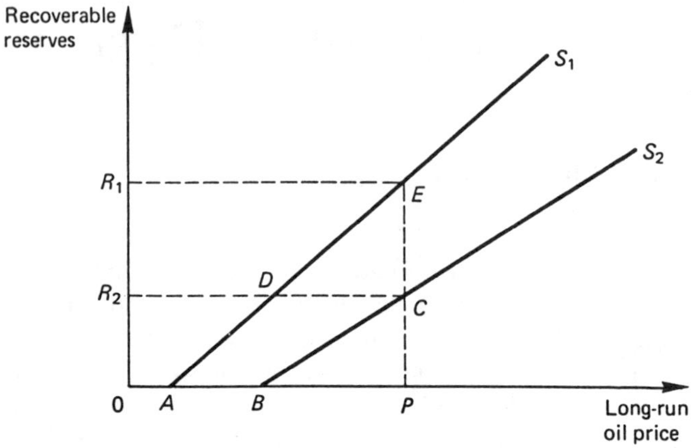

marginal profitability. This impact takes two routes. If marginal profits are affected by taxation this affects the supply schedule; first, via the impact on the normal return to capital in the margin, and second, via its impact on risk premia in the margin. By driving a wedge between the pre-tax and post-tax marginal returns and risk premia, taxation will reduce the willingness to supply. Both these routes are incorporated in our model. There is a third route, however, which will not be considered by our model. Taxation, by reducing *total* profits, may affect marginal risk premia (as explained in section 3.3). The direction of this effect is not unambiguous; however, we have argued earlier that it is more likely to exaggerate than to alleviate the sensitivity of supplies to taxation. That is, our approach may understate the scale of the impact of taxation on recoverable reserves.

Supply Curves and Producer Surplus

Economic theory suggests that if we can construct a supply curve we can also quantify the producer surplus as the area under an estimated supply curve.[1] Shifts in the supply curve will then imply gains or losses in the producer surplus, thereby providing some measures of welfare loss caused by taxation. For example, consider Figure 4.1. S_1 may represent the supply curve in the absence of taxation. If the tax shifts the supply curve from S_1 to S_2, at a price of P recoverable reserves will drop from R_1 to R_2 since the demand curve is assumed to

Figure 4.2 UKCS oil supply schedules (central case estimates)

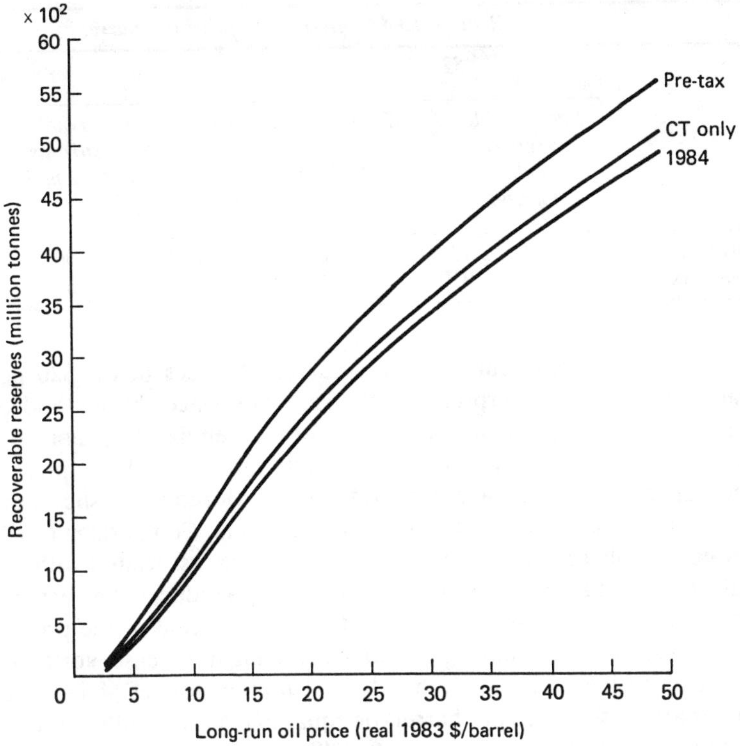

be perfectly elastic. This is a realistic assumption given the ease of substitution of foreign crude oil for UK crude oil. The producer surplus available to the private sector before any tax on pure profits will fall from *AEP* to *BCP*, while the government will benefit by *ADCB* as well as receiving a share of pure profits. The total producer surplus, however, will fall from *AEP* to *ADCP* and the loss in economic welfare may be measured by *CDE*.

4.2 COMPARATIVE STATICS ANALYSIS OF NORTH SEA PROFITS TAXATION

Figure 4.2 plots supply curves over the range of long-term oil prices from \$2.50 per barrel to \$50 per barrel (1983 values) estimated under

Table 4.1 Estimated recoverable reserves and the total surplus from offshore production given different tax systems

	Real (1983 $) long-run oil price per barrel			
	$20		*$40*	
	Recoverable reserves (million tonnes)	*Total surplus (1983 £ billion)*	*Recoverable reserves (million tonnes)*	*Total surplus (1983 £ billion)*
No taxation	2 798	90	4 837	364
1984 tax system	2 310	86	4 213	354

a variety of tax structures. Supply curves at prices below $20 per barrel need to be interpreted with great care since the method of analysis is purely forward looking (see Appendix B). Table 4.1 repeats two points of the range in absolute terms and also expresses the impact as a percentage of pre-tax levels. Moreover, it shows the size of the total producer surplus at these points. Comparison of the pre-tax supply schedule with the supply schedule associated with the current 1984 tax system suggests that taxation reduces the recoverable reserves to between 83 per cent and 87 per cent of the pre-tax estimates. That is, removing the tax system might increase North Sea oil recoverable reserves by nearly 500 million tonnes if long-run prices are close to $20 per barrel, or equal to over 620 million tonnes if long-run oil prices are close to $40 per barrel. The impact of taxation on the total producer surplus is less significant than the impact in quantity terms, but nevertheless it is still substantial.

Within the total impact of current taxation, the incremental effect of the two major taxes (CT and PRT) has been examined separately. CT alone has a considerable impact on the supply curve, accounting for about two-thirds of the difference between the pre-tax estimates and the estimates with the current tax system.

Examining the impact of the components of PRT in Figure 4.3 highlights the relative importance of the different PRT provisions. Tapering appears to be the most significant allowance. Without tapering, PRT would shift the supply curve about three or four times as far as PRT with tapering shifts the supply curve. At an oil price of $40 per barrel, supplies would be around 3680 million tonnes without tapering, while they are likely to be just 4213 million tonnes with the current tax system and just 4373 million tonnes with CT only. The

Figure 4.3 UKCS oil supply schedules (central case estimates)

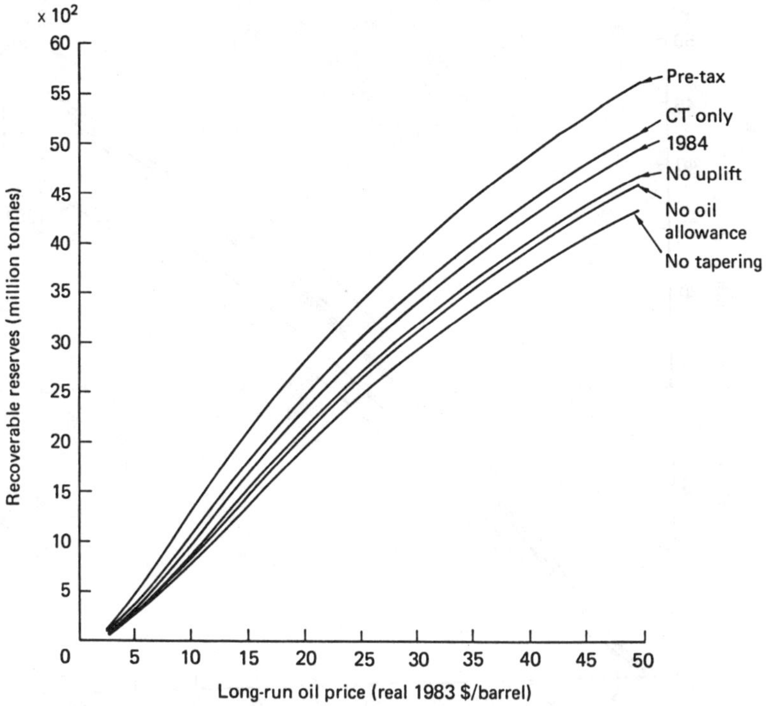

next most beneficial clause after tapering is the oil allowance. Of the three provisions considered, the uplift is the least advantageous. At $40 per barrel, the incremental effect of the uplift provisions is to add 240 million tonnes to recoverable reserves, while the oil allowance adds 318 million tonnes and the tapering and safeguard clauses add 535 million tonnes. One finding of Chapter 3 was that the impact of the uplift provision appears to be more relevant for intra–marginal fields than for marginal fields. Even the uplift, however, makes a significant difference to the supply schedule.

4.3 DYNAMICS OF TAXATION

Tax changes over time have altered the commercial supply prices and hence the long-run supply schedules. Figure 4.4 depicts the supply schedules for each of the seven tax systems between 1975 and 1984. These tax systems are listed in Table A.1 in Appendix A.

Figure 4.4 The impact of taxation dynamics on the supply of UKCS oil

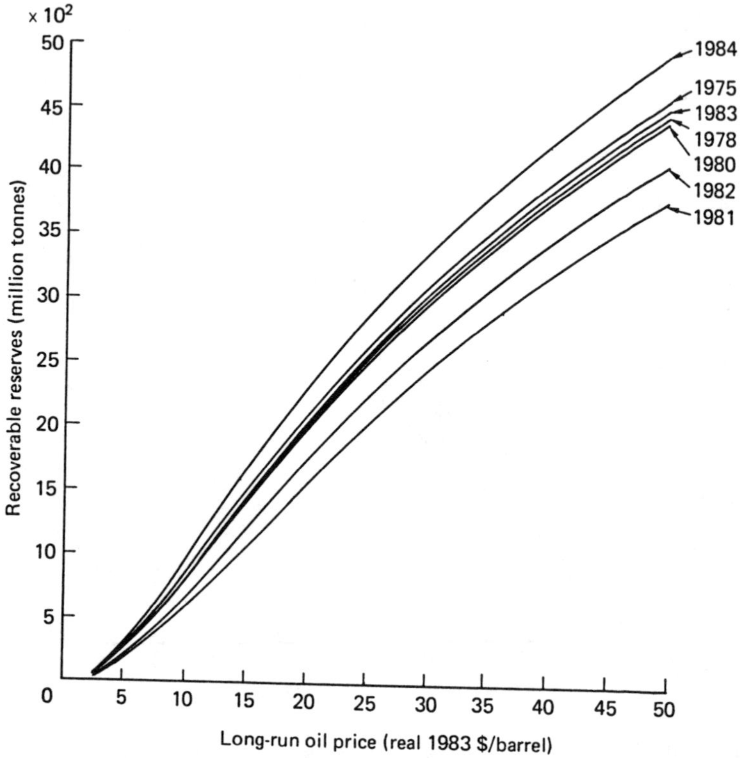

Note: Each date refers to the system of oil taxation introduced in that year.

With a long-run oil price of $20 per barrel, our model suggests the tax changes have reduced recoverable reserves by just under 500 million tonnes, from about 2800 million tonnes to about 2300 million tonnes. The biggest single shift came as a result of the third group of tax changes in 1981 with the fall in recoverable reserves amounting to over 420 million tonnes. The 1982 tax changes offset these reductions, adding about 175 million tonnes to recoverable reserves.

Chapter 3 argued that tax changes are not unrelated to long-run oil price expectations. If long-run oil price expectations rise then increases in tax rates or reductions in tax allowances are likely. The interpretation of Figure 4.4 in the previous paragraph may need to be adapted: only a curve such as the 1981 schedule should be considered with long-run real oil prices of $30 per barrel and only long-run real

Figure 4.5 Extrapolating the impact of taxation on the supply of UKCS oil

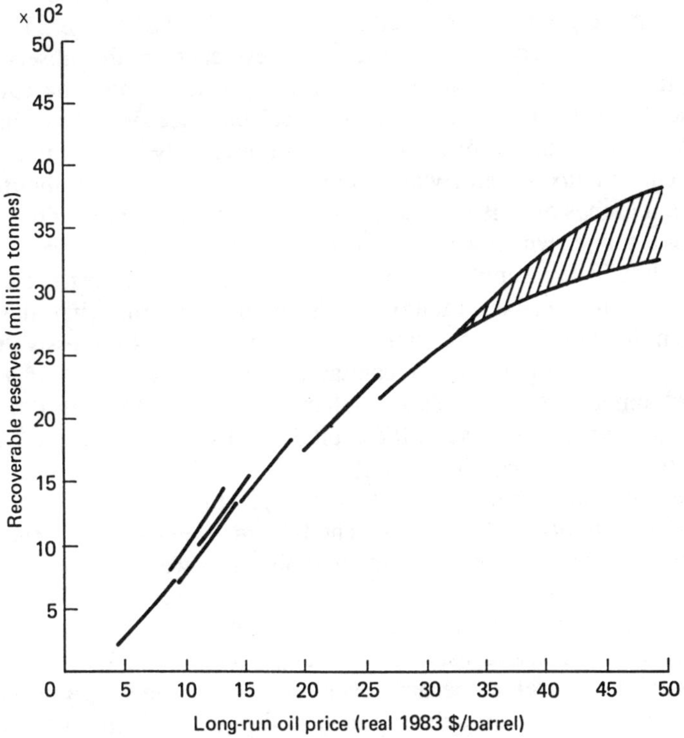

oil prices of under $10 per barrel should be considered with the 1975 schedule. Hence the long-run supply schedule is made up of a succession of segments of individual supply schedules associated with difference tax routines (which in turn are associated with a different long-run oil prices). This has been done in Figure 4.5 on the basis of the expectations brought together in Table 3.6. Extrapolating from these segments of the schedules suggests recoverable reserves about $30 per barrel might fall into the shaded area of Figure 4.5.

4.4 HAVE RECENT TAX CHANGES RESOLVED THE PROBLEMS OF TAXATION?

The spate of tax changes since 1982 appear to have improved the taxation of marginal fields and hence to have increased the level of recoverable reserves, but serious problems remain. First, although the supply curve shifts in the correct direction (that is, towards the pre-tax schedule), the scale of the shift is small relative to the distance between the pre-tax supply schedules before the 1982 budget. That is, considerable reserves which would be developed in the absence of taxation are not likely to be commercially viable under the current structure of taxation. At a long-run real oil price of $20 per barrel some 2310 million tonnes would be commercially recoverable with the current tax system, which represents an increase of about 578 million tonnes over 1982 expectations but is still well short of the 2798 million tonnes which would be recovered in the absence of taxation. The shift in the supply schedule does not bring the post-tax curve close to the pre-tax situation. This finding confirms that the tax system for new North Sea fields, although made less regressive than formerly, is still a tax on capital and is still seriously affecting development prospects. This problem was worsened by the 1984 CT changes, largely because of the capital intensive nature of the North Sea oil industry (see section 3.3).

Second, the dynamics of taxation imply the scale of the tax reduction will bring the supply schedule in line with the schedule pertaining when expectations for real oil prices were similar to those held currently. For example, the recent decline in oil prices has led to expectations of constant nominal oil prices for perhaps as long as 10 to 15 years which suggests a real long-run oil price in 1983 values of $10–$15 per barrel, or perhaps even less. Hence the supply schedule is likely to shift towards the segment of the curve which portrays the dynamics of taxation applying between $10 and $15 per barrel. Indeed, the impact of the tax changes is almost exactly predicted by the curve portraying the dynamics of taxation and our earlier reasoning is confirmed. Moreover, the area of the supply schedule predicted for long-run oil prices above $15 per barrel remains largely unaffected. This is due to the relationship between oil price increases and tax changes (See section 3.4).

In short, despite what appears to be favourable tax changes since 1982, the scale of the damage caused by North Sea taxation to the supply of UKCS recoverable oil reserves remains considerable.

4.5 CONCLUSION

Our comparative statics analysis of taxation suggests the tax system might reduce UK ultimately recoverable oil reserves by between 488 million tonnes and 624 million tonnes. The dynamics of oil taxation may further reduce UK ultimately recoverable oil reserves if real long-run oil prices reach $40/bbl by around 1000 million tonnes. In terms of the lost economic welfare, the system may decrease the total producer surplus in 1983 values by between $4 billion and $10 billion, while the tax dynamics may be responsible for an additional fall of similar magnitude.

The size of the impact of taxation is substantial. However, removal of the tax system, rather than its replacement, is not feasible. The decline in government revenues would not be tolerable to politicians in government. Clearly though, there is room for improvement in the tax system, and surely a replacement exists which would leave government revenues barely altered but would not have the same impact on marginal profitability and would not shift the supply schedule as much as current tax policy.

The next chapter discusses one possible replacement.

5 Petroleum Profits Tax: An Alternative Proposal

5.1 AN ALTERNATIVE TO CURRENT TAX POLICY

Before we can finally condemn the present tax system we must introduce an alternative that has less of an impact on development decisions but which still satisfies government revenue requirements. Analysis of the Institute for Fiscal Studies (IFS) proposals[1] (below) suggests that these recommendations represent such an alternative where the costs of invention, legislation and publication can be kept to a minimum. Clearly, however, the IFS proposals were not seen to be an obvious improvement to all since they were not implemented when first published at the end of 1981.[2] By asking why they were not implemented,[3] though, the false perceptions and the remaining costs that prohibit its implementation can be ascertained. This we do in the next chapter; our purpose here is to establish the merits of the IFS proposals.

This chapter argues that, had the IFS proposals been implemented, neither the industry nor the government would have been worse off, and both would probably have been made better off compared to the existing system of taxation. In short, the IFS Committee recommended that the contemporary tax regime of royalties and PRT be abolished and that CT no longer apply to North Sea oil operations.[4] In its place, Petroleum Profits Tax (PPT) would be levied. As explained in Chapter 2, pure profits in the offshore oil industry cannot be eliminated by competition, an oil field cannot be reproduced and hence each oil field should be considered as a separate profit unit. For this reason PPT was constructed as a field-by-field tax and there is no convincing reason why exploration costs should be offsettable against current PRT; instead exploration costs should be rolled forward (and compounded) as costs of new projects. PPT is based on profits compounded at three rates of interest to give three bands of tax.

The rates of interest are related to real rates of return on investment. The three threshold rates of interest, below which no tax in each band is payable, would be set at 15 per cent, 25 per cent and 35 per cent, with corresponding tax rates of 50 per cent, 25 per cent and 10 per cent. This implies that until a 15 per cent rate of return is reached no tax will be payable; between rates of return of 15 per cent and 25 per cent profits be taxed at a rate of 50 per cent; between rates of return of 25 per cent and 35 per cent an additional tax of 25 per cent will be payable (giving a total rate of 50 per cent + 25 per cent = 75 per cent and for profits above 35 per cent an additional tax of 10 per cent is levied (giving a total of 85 per cent). Under the IFS proposals, fields would be subject to PPT over their entire life and the existing tax system is no longer relevant.

Chapter 3 explained the weaknesses of the current tax system and made suggestions for its improvement. On the one hand, we discovered that the provisions in PRT do not perform the function we would like them to perform and, on the other hand, we found that the relationship between PRT and CT impedes the proper functioning of the tax system and the distribution of its burden. Improvements would then fall into one of two categories: either simplifying PRT or simplifying the interconnections between PRT and other taxes. Since the IFS proposals represent both types of simplification, there are strong reasons for believing it will be an improvement.

The impact of implementing these proposals is now assessed, first on the government's position via a comparison of the aggregate tax revenues likely under the different tax systems and then on the industry via an examination of inter-field profitability. Finally, we assess the impact on the supply curve.

5.2 THE RESOURCE RENT TAX

A tax system ought to extract the economic rent – and only the economic rent – from an industry because the producer's decision-making process would not then be affected by the tax. That is, in the oil sector, the pace of production, the technique of extraction and the choice between competing projects would not be distorted by a government that captures economic rent via the tax system.

In principle, there are two ways in which the economic rent may be extracted.[5] First, some system of competitive cash auctioning of oil licenses (including all rights of extraction) could be used to stimulate each company or consortium to bid their valuation of the expected economic rent. As a result, it would be reasonable to expect the lowest cost producer would win the licence[6] (see section 2.2) and, providing sufficient competition existed amongst companies or consortia, the full value of the economic rent would accrue to the licensing authority and no further taxation would be necessary. There are two main drawbacks to this procedure. First, the auction must be truly competitive, otherwise not all the rent will be extracted. Second, as it is an *ex ante* method of rent extraction, there is always the danger that at a late stage some further tax may be imposed in response to political pressures or changed economic expectations. This danger will itself depress the level of bids in an auction.

The second method of capturing the economic rent is by some form of *ex post* tax. Taxes, however, tend to distort the decision-making process, particularly when applied to long lead time extraction industries. An awareness of these problems led economists to search for a tax that would be neutral when applied to long lead time extraction industries such as the mining industries, originally, but subsequently to the offshore oil sector. The earliest proponent of such a tax is generally considered to be E. Cary Brown in 1948.[7] The most innovative features of the so-called Brown Tax, based on discounted company cash flows, are the full loss offset available and the rapid depreciation of assets. Because the full loss offset means that capital costs in each period are effectively reduced by the rate of the tax, and as net incomes in each period are also reduced by the rate of tax, then the project's rate of return would be the same before and after the imposition of the Brown Tax.[8]

In 1975 Garnaut and Clunies Ross[9] developed a modified version of the Brown Tax specifically for natural resource projects in those Less Developed Countries where the finance is provided externally. In these instances, the funds employed do not affect the rate of return available on competing investment opportunities. This resource rent tax (RRT) is based on cash flows after losses have been carried forward at the rate of interest, known as the 'threshold rate'. The main difference[10] between RRT and the Brown Tax is that RRT involves carrying forward losses at some rate of interest, whereas the Brown Tax provides a rebate for losses. Advantages of RRT over other systems of taxing natural resource projects[11] include RRT's ability to respond automatically and progressively to a variety of outcomes, its avoidance of the problems associated with forecasting costs and prices since it is based on revealed profitability, and its lack of distortions to investment and production decisions (its neutrality).

Garnaut and Clunies Ross[12] define a neutral tax as one which does not alter the order of competing investment projects. For RRT to be neutral then, the threshold rate should reflect the investor's discount rate. Of course, the investor's discount rate will itself vary between investors, over time and between projects. In the initial structure of RRT, because the finance is external and for a specific projects, the capital is not being diverted from another project in that country and hence the issue of neutrality among industries is not a concern. In an industrialised country neutrality among industries must be considered. Thus to minimise distortion, threshold and tax rates ideally should be developed for individual circumstances with RRT accrued on a project rather than a corporate basis. To resolve these problems in practice, Garnaut and Clunies Ross[13] suggest progressive scales of RRT rates based on various threshold rates for carrying forward losses.

Under certain circumstances RRT may have a distortionary effect on capital investment and on the rate of resource depletion.[14] However, these problems are associated with the issue of how to define – precisely and practically – profitability. Various measures of profitability have been recommended[15] and the principle of a resource rent tax is not undermined. Indeed, in only a few cases would the problem of measuring profitability affect the functioning of a rent tax in the UK North Sea.

In natural resource projects characterised by long lead times and high capital front-end loading, risk is a crucial issue over which there is disagreement as to how it should be considered in the tax system. Sumner[16] argues that a certain positive level of NPV should be required before the project is undertaken. Clunies Ross,[17] on the other hand, argues that as companies generally account for risk by raising the discount rate, then so the tax and the threshold interest rates should be increased. However, it is the investors (shareholders) rather than companies who are taking the risk and investors will not raise the discount rates to take account of risk; rather they will diversify their investment portfolios. Moreover, it is doubtful whether companies arbitrarily raise their discount rates in response to risky ventures.

A feature of the existing North Sea tax system is the possibility that wasteful capital expenditure will be encouraged (gold-plating). This occurs if sufficiently more than 100 per cent of incremental capital expenditure can be offset against tax payments to compensate for the interest lost on the time delay between making the capital expenditure and reducing the tax payments. RRT would not eliminate this possibility but it would make gold-plating much less likely to occur. As long as the rate at which losses are accumulated for the assessment under RRT closely approximates investors' discount rates, it would be difficult to increase a project's NPV by incurring unproductive expenditure and the incentive for wasteful capital spending will be minimised.

Palmer[18] maintains that RRT may be politically unacceptable as it may delay tax payments and hence can only be imposed in conjunction with CT. This is a problem which the IFS report found was not relevant to North Sea taxation in the 1980s since the period of tax holiday had already passed. Indeed, in switching over to PPT, the IFS observed a close approximation in both timing and the magnitude of tax payments that would occur under PPT compared to projections of tax revenues under the existing tax regime, so that further taxes would not be required at that relatively advanced stage in the development of the UK North Sea. In addition, the IFS system would be internationally compatible (specifically with the US), thus further diminishing Palmer's doubt. It is worth noting Clunies Ross's approval of the IFS tax proposals:

> This system would be extremely simple, it would be responsive enough to profit to be very probably stable; it would, with some qualifications, approach neutrality in the most important respects; it would be unimpeachable on grounds of fairness; and the Committee show . . . that it could be expected to raise revenue comparable on average to that raised by the existing system on each field exploited.[19]

5.3 IMPACT ON GOVERNMENT

The government appears to have several criteria in its requirement for North Sea taxes. The criterion with the highest priority is probably the scale of early tax revenues. If early revenues fall short of the level expected under the existing tax system, the impact on the PSBR may be significant enough to force government to alter its macroeconomic strategy. Any tax system which failed to produce a tax yield similar to the current tax system would be unacceptable.

Another criterion for government North Sea policy relates to more general principles of political strength. It is in the government's interests to be seen in charge of the oil sectors, particularly in times when there is considerable public awareness of events in the oil industry; for instance, when oil prices rise. This suggests a system which increases the tax take when oil prices are high – such that tax revenues increase more than proportionately to pre-tax profits – would be favoured.[20]

The total tax take calculated over the lifetime of North Sea oil is another criterion affecting government's attitudes to North Sea tax. The government desires a fair tax take and what constitutes 'fairness' is largely a political decision.

These criteria all relate to aggregate tax revenues. The Chancellor emphasised this concern at the start of 1981 when he called for suggestions to change the system of taxation.[21] He invited proposals which would not significantly affect total tax revenues and declared that all such proposals would be given serious consideration. Conversely, the field-by-field distribution of the tax burden does not seem to be an area receiving high government priority.

The criteria identified above for the government may not be government targets *per se*, but probably stem from other, implicit, criteria relating to the personal motivation of politicans and civil servants. It may not, however, be possible to express some of these implicit criteria in terms of the three criteria relating to aggregate tax revenues described above, especially if the implicit criteria can be fulfilled by a certain type of tax system only. That is, if the means (a tax system) of achieving a given end (a total level of tax revenues) is important then the criteria listed above will not be sufficient to ensure all public sector agents are no worse off. An example of a situation where the means are more important than the end would be the preference given by tax authorities to more complex tax systems over more straightforward tax systems, because more complex tax systems give the tax authorities more kudos and influence. An analysis of how the IFS proposals relate to preferences for the means of tax policy is, however, delayed until it is established that the ends of tax policy are satisfied by the IFS proposals. Chapter 6 returns to the issue of how the IFS proposals affect preferences over the means of taxation.

Aggregate Tax Revenues

The IFS Report maintains that total tax revenues from oil fields under the existing system and with PPT are remarkably similar, and we are able to confirm this. The IFS proposals satisfy the first government priority of a sufficient take.

In addition, with a low oil price forecast, the IFS proposals are likely to produce a little more in aggregate tax revenues than the existing tax system. The bulk of additional revenues would occur later, so that discounted tax revenues would not be much affected. With a relatively high oil price, PPT collects significantly more than the existing system. Again the bulk of additional revenues occurs later, which would be expected as it is the tax treatment of the fields not yet under development which alters most. PPT would be likely to improve the sensitivity of the tax revenues to oil prices and because PPT acts progressively with respect to oil prices, stability in the system is promoted.

5.4 INTER-FIELD BURDEN OF TAXATION

In principle, PPT fulfils many of the lessons discussed in Chapter 3. PPT attempts to achieve progressivity not by arbitrary concessions or allowances but instead via a rising schedule of tax rates.

The high tax rates will be levied where profits exceed prescribed rates of return: a better measure of field profitability than field size as suggested by the oil allowance in PRT. Thus with PPT, the average tax take is likely to be positively related to field-by-field profitability. The tax allowance for capital spending automatically reflects both each field's exposure to front-end loading of costs and the interest burden implied by such front-end loading. Indeed, the IFS recommend that changes in either real interest rates or inflation which lead to a significant movement in nominal interest rates should be matched by altering the threshold rates, thereby keeping the normal returns to capital protected. By constructing a tax base which conceptually matches pure profits, available risk premia in the margin will not be affected by taxation.

As indicated above, PPT is likely to be automatically sensitive to oil prices and hence the need to increase taxation in order to increase the aggregate tax take will not be present. This should not only remove the desire to tax incentives but should also promote stability. If, however, government still wishes to be seen to be in control of the international oil companies operating in the UK North Sea, PPT offers a simple and effective route for government intervention through the scheduling of tax rates. Although uncertainty as to when tax changes may arise will not be eliminated, at least uncertainty as to what tax change may occur would not hold back development decisions. Furthermore, since the PPT tax base resembles pure profits more closely than the tax base under the current tax system, then tax instability will not aggravate other defects in the tax system.

For those fields already granted development consent by the Department of Energy, the inter-field tax burden is shifted towards the more profitable fields. Comparing the post-tax NPVs per tonne with PPT to post-tax NPVs per tonne available under the existing system, it appears that most fields would earn greater profits if the IFS proposals were adopted. With respect to fields yet to be granted development consent, three of the most profitable projects (Columba, Scapa and Montrose South) have lower profits. This comparison assumes no CT can be deferred by the development of new fields. If there are ample CT payments which can be deferred under the existing tax system the comparison implies that for the most profitable fields only about half are better off (T-Block, Crawford, Lyell, Eider and Clair). It is noticeable that whatever the standard of comparison, the IFS proposals consistently improve the profitability of the less attractive new fields (yet to be granted development consent) whose development is most endangered by the present tax system.

We now compare the average and marginal government takes under the existing tax system to those under the IFS proposals. The change to PPT would introduce considerable progressivity into average government takes where it was lacking under the existing system. Given progressive average government takes, it is no longer so desirable to have progressive marginal tax takes.[22] Indeed, for a system that has a top rate of tax of 85 per cent which is asymptotically approached from below as profits increase, the marginal rate of tax will be inversely related to the average rate of profit (Figure 5.1). The desirable progressive nature of PPT is emphasised by average tax takes (in nominal terms) rising towards the asymptote of 85 per cent while nominal tax takes fall towards the asymptote.

Incidentally, marginal tax takes (in nominal terms) of over 100 per cent under PPT need not concern us. This is an undiscounted value and does not necessarily imply wasteful capital spending will be encouraged. For wasteful capital expenditure to be encouraged, marginal real discounted profits must rise with unproductive capital expenditure (this is the problem of gold-plating). Marginal nominal government tax takes of more than 100 per cent imply that undiscounted nominal profits might rise with unproductive capital expenditure. This occurs because greater undiscounted profits may reduce the undiscounted value of the capital deduction by bringing forward the date when tax thresholds are crossed, and thus increase the undiscounted tax base more than proportionately to the increase in

Figure 5.1

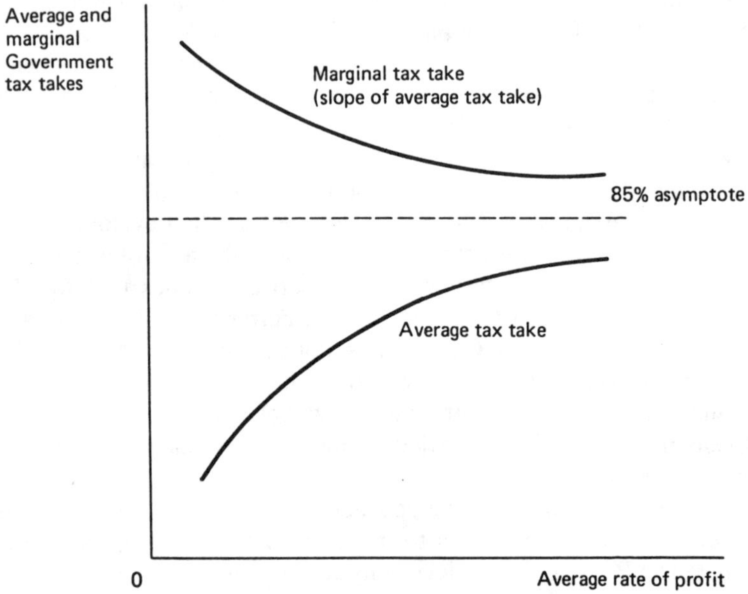

undiscounted profits. The size of the marginal tax takes will be greater the greater is the reduction in the period before tax thresholds are met. However, the rate of return available to those considering wasteful capital expenditure will be limited by the threshold rates of return in PPT. In the limit a maximum nominal annual rate of return of 35 per cent would be available. This limit, however, is approached very slowly as the length of lag approaches infinity: it remains below 20 per cent even over a ten-year period.[23] As this is expressed in nominal terms it does not seem to be a strong enough incentive for investors. Moreover, when expressed in real terms, the marginal government taxes are never above 100 per cent. That is, it would not be possible to achieve a positive real rate of return by wasteful capital spending. This is not true under the existing system.[24] In short, the problem of gold-plating which occurs in the existing tax system is not resolved by PPT; however, PPT does not aggravate the problem.

PPT is able to distinguish between the profitable and the less profitable fields, and to be sensitive to the factors which affect the attractiveness of North Sea investment on a field-by-field basis. Average government takes tend to be closely related to field profitability and fall or rise automatically with unfavourable or favourable oil market conditions (that is, lower or higher oil prices). Again, because the changes are automatic, PPT would not tax incentives. A problem with respect to taxing incentives is that governments change average tax takes; under PPT the only additional burden would be on the marginal take.

The inter-field tax burden under PPT seems to shift the burden of taxation away from the marginally commercial fields which are adversely affected under the existing system. It seems probable that, if the IFS proposals were adopted, North Sea investment would appear more attractive in the margin, more fields would be developed, the period of net oil exports would be lengthened and the balance of payments would be boosted.

5.5 IMPACT ON RECOVERABLE RESERVES

Figure 5.2 portrays our central case estimates of the effect of existing tax policy on recoverable reserves (see Chapter 4). By re-estimating the supply equation (as described in Appendix B) the impact of changing to the IFS proposals can also be assessed. We find the commercial supply oil prices tend to be significantly reduced, so that re-estimation of the commercial supply equation yields much higher reserves.

Using PPT in our behavioural model suggests UK recoverable oil reserves would be substantially increased (the dashed line in Figure 5.2). The size of the increase depends on the long-run real oil price assumed. With a real $20 per barrel (1983 values) long-run oil price, our model suggests ultimate recovered oil reserves would increase by 217 million tonnes. At a real oil price of $40 per barrel, the increase from switching to the IFS proposals would be as great as 315 million tonnes.

Similarly, the lost total surplus form the UKCS would be much reduced under the IFS system. The total surplus under the IFS supply schedule is almost £88 billion (1983 values) at a real $20 per barrel (1983 values) long-run oil price and is about £360 billion at a real price of $40 per barrel, compared to the 'no tax' total surplus at these points of £90 billion and £364 billion respectively. The potential increase in surplus realised by switching away from current oil tax policy and towards the IFS proposals would be between £2 billion and £6 billion (real 1983 values).

Figure 5.2 Various tax structures and the supply of UKCS oil

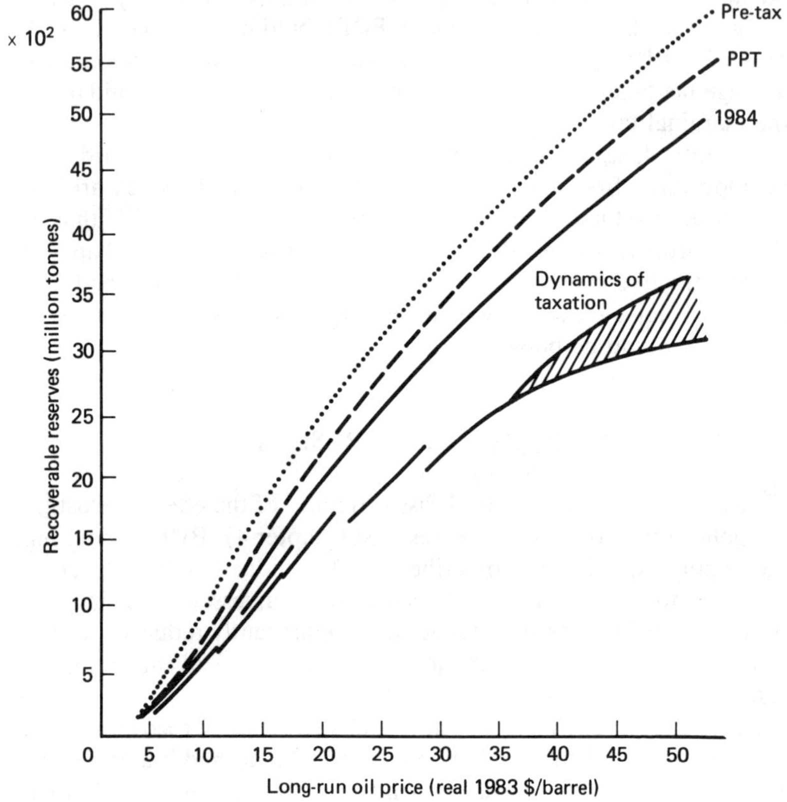

5.6 CONCLUSION

We suggest above that government revenues would become more sensitive to oil prices if the IFS proposals were adopted. Thus some of the pressure to amend tax rules whenever oil prices change would be relieved. Whether the added sensitivity would provide sufficient additional tax revenues is difficult to ascertain. In any case, this is not a reaction based solely on raising extra revenue. It may also be a response to the electorate's view of the oil industry's capacity to bear taxes. Changing to the IFS proposals may not alleviate all the pressure on the dynamics of taxation and other changes may still occur. The IFS report was aware of this pressure and specifically recommended that if changes have to be made they should be made to the tax rates, not to the tax bands, in PPT.[25] Consequently, although the dynamics of taxation may persist, the adverse effects of the dynamics of taxation would not be so severe.

In terms of the supply schedule, the impact of tax dynamics under PPT would be much reduced compared to the impact of the tax dynamics under the current tax system. At any given oil price scenario, the range of recoverable reserves is very narrow compared to the range of recoverable reserves likely with current tax policy. The effect of a less extreme set of rates of PPT on aggregate tax revenues under an assumption of high oil prices and considerable exploration activity indicates that the aggregate tax revenues are sensitive to PPT rate changes even though the supply schedule is not greatly disturbed.

The impact of tax dynamics would be much reduced by moving to PPT. Not only would there be reduced pressure for tax changes, but tax changes would not be as significant as with the current tax system. This result emphasises how much closer the tax base in PPT is to economic rent than the tax base in the current system; thus PPT would significantly reduce the distortionary impact on the offshore oil industry.

6 A Desirable Solution

6.1 A MISTAKEN POLICY

Had the IFS proposals been implemented in 1982, the government (in terms of end criteria) would have been no worse off, the inter-field burden of taxation would have been more appropriate and significant additional reserves would have been realised. An improvement for all would, apparently, have been inevitable. Clearly though, it was not seen as an improvement by all; but why not?

Our interpretation of the events which culminated in the rejection of the IFS proposals is that the government appeared to be willing to implement the changes if the IFS proposals received industry backing. This backing was not forthcoming for several reasons which still tend to persist. However, the onus now must be on the government to push for change.

6.2 EXTENT OF DAMAGE

First, a full appreciation of the extent of the damage attributable to the existing tax system did not (and does not) appear to be common. Hence an appreciation of the extent of possible improvement was (and is) also lacking. Preceding chapters have shown the extent of damage to be considerable and the scope for improvement to be great. In particular, Chapter 3 outlined the distortionary impact on resource allocation over time caused by the tax system, and Chapter 5 estimated the improvement to the supply base which would accrue as a result of implementing the IFS proposals.

6.3 HIDDEN COSTS

Second, perhaps our analysis has omitted to consider some of the costs in changing to PPT. If some economic agents bear costs without any compensating benefits, and are thus made worse off by the changes, then the claim to be an improvement to all would be mistaken.

Bureaucratic Needs in the Public Sector

Within the government bureaucracy, such costs may be costs of administrating the new legislation or costs in loss of prestige, power, influence, employment, or other non-financial variables affecting bureaucratic utility. The administrative workload for running PPT would not, however, be greatly different from the workload of PRT – the biggest administrative workload is in agreeing field boundaries and supervising cost claims – while the required legislative changes of establishing the IFS proposals would not be particularly onerous.[1] Although the IFS proposals are much simpler to understand, no loss of importance would be felt by the bureaucrats. Preference over the *means* of raising tax revenues would not be much affected by the switch to PPT.

Perhaps the bureaucrats might feel worse off because they were not responsible for devising the new scheme, whereas they may be held responsible for the faults of the existing system. Thus their standing and credibility within the industry would suffer. This may occur but it would not be a consequence of adopting the IFS proposals. In the discussions leading up to the March 1982 budget,[2] it was accepted and recognised by all that PPT would not have been possible in 1975. The front-end loading of allowances in PPT would have delayed early tax payments and sufficient early tax revenues would not have been forthcoming: hence one of the government's criteria would have been unfulfilled. Thus implementing PPT now would not be an indictment of the past performance of the existing tax system, only a criticism of the existing tax system's future performance. Furthermore, the faults of the existing system had been declared by the Chancellor when he invited proposals for a different tax regime.[3] The IFS recommendations did not uncover any previously unknown faults and did not threaten the credibility of bureaucrats.

Perhaps the conceptual simplicity inherent in PPT would be unattractive to bureaucrats since this may reduce the number of occasions when they are called on to give advice to Ministers. Although we doubt that Ministers would seek fewer consultations, even if this occurred the role of the bureaucracy would probably still be increased as the stimulus to North Sea activity would create additional demands. Although some loss in psychic income derived from consideration of the prospects for any one field may follow from PPT's conceptual simplicity, the increase in activity would lead to an increase in the number of fields to be considered.

Some transfer of workload within the bureaucracy may be necessary, but given the scale of additional activity this transfer should be possible without leaving anyone worse off.

Bureaucratic Needs in the Private Sector

By a similar line of reasoning, the specialist tax advisors of oil companies might not have welcomed the IFS proposals. Simplifying the tax system can be quoted as an advantage to company planning, but company tax experts might not welcome it as their specialist skills developed in understanding and manipulating the complicated and cumbersome existing tax system would become redundant. As above, though, perhaps the specialists would have a reduced workload in analysing the financial prospects of any given project, and perhaps redefinition of some specialist roles might be necessary, but with the additional activity likely under PPT, an overall increased workload (and importance) would be likely.

Hence fears of tax specialists in both the public and private sectors were probably unfounded. They might have resisted change because they perceived a loss of their influence and power although such a loss would, in fact, have been unlikely. Apparently the tax specialists ignored the stimulating influence of the IFS proposals on North Sea activity; Chapter 5, though, indicated that the stimulus would have been significant. Thus they foresaw a reduction in their kudos for any given project but failed to forecast an increasing number of projects. Such false perceptions may have created substantial resistance to a change that would probably not have been detrimental.

Intra-Industry Effects on Portfolio Values

When considering future potential gains we may be justified in assuming that what benefits the private sector will benefit all companies, but when considering the impact on existing portfolios of assets we must be aware of differences between companies. If some companies suffer greatly from the changes then even net of future gains they may be worse off. If there were only small future gains to be earned by all, many of the companies which are likely to lose on assets already in production or under development would also lose net of future gains. Compensation would then be necessary for the net losses, otherwise PPT would not be an improvement for all. This compensation might be extraordinarily costly. Negotiating compensation between companies would require explicit agreement on the future impact of PPT, an impact which would be based on expectations of a range of imponderable parameters.[4] Companies paying compensation might fear some precedent would be set, anti-trust laws might be infringed and a long-term more aggressive government policy might be anticipated; all such factors would make the costs of facilitating compensation substantial.

Fortunately, compensation is not a necessary requirement since potential future gains are not small. Although some companies might lose on their existing developments, since many more of their discoveries would be commercial they would be likely to gain overall. Resistance based on the impact on existing developments only may help to explain why the IFS proposals were rejected, but such resistance would be based on incomplete and therefore false perceptions.

In any case, our ultimate concern here should be for investors in a company, not for the company *per se*, and investors are likely to have spread their portfolios across many companies. This would reduce the likelihood that any individual would suffer.

Costs Relative to Potential Gains

We have suggested above reasons why individuals in the public or private sectors may be (and might have been) wary of changing to the IFS recommendations. We also suggested reasons why such perceptions may be false. In particular, a proper appreciation of the extent of future possible gains makes these costs seem insignificant. Hence, there is (and was) misconceived resistance to the IFS proposals.

6.4 PRIVATE SECTOR PERCEPTION OF GOVERNMENT REQUIREMENTS

A third explanation for why the IFS proposals lack (or lacked) support is now considered. We believe the private sector's perception of the government's requirements for tax revenue is (and was) inaccurate.

Industry Perceptions

The industry appears to believe the government would be willing to accept a decline in its tax revenues from the North Sea. Both bodies representing the industry – UK Offshore Operators Association

(UKOOA) and The Association of British Independent Oil Exploration Companies (Brindex) – argued in 1981[5] that the total tax burden is excessive and must be reduced in order to stimulate North Sea activity.

> UKOOA's conclusion is that taxes of the magnitude described are excessive and therefore believes that the Chancellor's objective of obtaining a 'broadly similar yield' is not realistic . . . that taxes are at present too high is evidenced by the fact that, since the 1981 Budget annoucement, several operators have publicly expressed serious concern about the viability of several proposed UKCS field developments and as a result the following projects are the subject of detailed reconsideration. Andrew, South Montrose, SW Claymore, Tern and Thistle (Area 6).[6]

> The Brindex proposals for a revised tax structure are accompanied by the submission, strongly felt by Brindex members, that the total government take by project and by company should be reconsidered in order to make possible an expanding British oil and gas industry.[7]

> Brindex believes the total amount of royalty and taxation under the present system is too high for the good of the British oil industry and militates against maximizing the wealth from this valuable national resource.[8]

Thus, one aim of the Brindex proposals was: 'a lowering of the tax take both to reduce the threshold for viable field development and to stimulate exploration'.[9]

A reduction in total taxation is seen to be feasible by the industry. UKOOA's quantification of their proposals shows they sought a reduction in aggregate tax revenues of more than £1 billion per year over the five years from 1981 compared to the system of taxation in force after the Budget of 1981, and a total reduction of between £11 billion and £39 billion over the life of North Sea oil.

Consequences of False Perceptions by Industry

So long as the industry believes in the feasibility of reducing total taxation, it will tend to argue that a reduction in total taxation is *necessary* to stimulate marginal field development. Total taxation refers to aggregate tax revenues and tax takes for the whole of the North Sea. It is largely determined by intra–marginal projects and is assessed by reference to average tax takes. On the other hand, marginal taxation refers to the impact of taxation on projects on the borders of commercial or economic viability. It is assessed by reference to the structure of the tax system. Since the industry stands to gain more by a reduction in total taxation than a reduction in marginal field taxation, it will tend to be more concerned about the total tax burden than the tax treatment of marginal fields.

The industry's bargaining strength, however, is related to new field developments, which tend to be marginal fields, rather than the existing projects where capital costs have been committed already, and which to a large extent determine the total level of taxation. It would be a hollow threat if the industry were to force the government into lowering the level of taxation by announcing plans to reduce production from existing fields. Its only threat is not to develop new fields. Hence, the industry claims that some fields will not be developed unless the level of taxation on existing fields is reduced. Clearly this threat is not made explicitly. Instead the industry argues that it is necessary to raise the funds for new developments from equity sources, that equity funds come from existing projects only, and that improving marginal prospects alone is not sufficient to create greater enthusiasm for North Sea investment.[10] Thus the case for more lenient treatment of marginal fields – a case which we strongly support – is invoked to justify the call for lower total taxation, a call whose relevance to promoting North Sea investment we doubt.

Consequently, the inter-field burden of taxation is of secondary importance to UKOOA. Their proposals would yield benefits to the industry by reducing average government tax takes on fields in production or under development but not by improving the structure of taxation, and the inter–field tax burden would not be improved. The prospects for potentially commercial fields would also be improved, but the improvement would be at least as great for the more profitable fields not in need of improvement as the improvement on the less attractive fields whose development is most in jeopardy.

To a large degree, the industry argument relies on some factor which inhibits marginal profits from directing investment funds to North Sea projects. This fact is often assumed to be a capital market constraint. The existence of such a constraint, however, is doubtful. Although companies may prefer to find new projects from North Sea equity sources, this is unlikely to be a binding constraint on North

Sea developments. North Sea companies have access to non-North Sea equity as well as debt market sources and non-North Sea companies are keen to enter the industry. Indeed the findings of the Wilson Committee on financing North Sea oil suggest the capital market has not constrained, and does not restrict North Sea investment.[11] Perhaps the level of equity funds does influence North Sea activity, but it is not a constraint.

Even if the industry realises the capital market constraint is illusory, it is likely to continue to argue the inextricable connection between the total tax burden and marginal field activity. Without this connection there would be limited possibilities of persuading government to reduce the total tax burden since this would be an issue of distribution solely. A step-by-step approach, where the marginal issue would be solved independently of the total level of taxation, could not be accepted by the industry as this would imply the two issues are separable. The IFS proposals, however, represent such a step-by-step procedure. Hence if the industry backed the IFS proposals it would lose the chance to influence the total level of taxation.

So long as a reduction in the total level of taxation is seen to be feasible, the IFS proposals will not appear to be an improvement on what might otherwise occur and will not receive industry backing. A lower total level of taxation would be preferable to the industry.

Moreover, arguing for a reduction in the total level of taxation may be an inevitable consequence of two factors. First, in order to represent the industry effectively, its bargaining organisation, UKOOA, needs to have unanimous support from the industry. Given the companies' different tax positions, it may only be possible to reach unanimous agreement by calling for a lower level of total taxation. Second, within the structure of the 1981 tax system when discussions were taking place, it was probably impossible to find any mechanism to reduce the tax burden on marginal fields which could not also reduce the total tax burden. Without radical changes, a reduction in the tax burden on marginal fields would necessarily also involve a reduction in the total level of taxation.[12]

More Realistic Attitudes

The arguments above suggest the industry perceives that a substantially lower level of total taxation would be acceptable to the government. This perception proved to be false in March 1982 and we believe it is still false. Once the industry realises the false nature of its belief a different outcome might occur. Accepting the constraint on total taxation, the industry might then welcome improvements in the marginal tax structure for their impact on the marginal fields *per se*, rather than looking for changes in the structure of taxation which would improve the tax position of intra–marginal fields. The IFS proposals would then be seen as an improvement to the industry over what might otherwise occur.

PPT would appear to be an improvement for all against what might otherwise prevail. On intra–marginal existing fields, the government would be satisfied by the level of tax revenues. The industry would realise this level of tax revenue on intra–marginal fields is a constraint imposed by policy and not an object of policy to be influenced. Prospects for marginal fields would be more attractive, North Sea activity would be stimulated and the total surplus from North Sea oil would increase. The division of the additional surplus would be between the government and the private sector. To the extent that the government receives some of the additional surplus it would be likely to prefer PPT over the existing tax system. The major beneficiary of the additional surplus, though, would probably be the private sector.[13] Our contention is that the size of the additional surplus would be so great that even if a company's portfolio of existing fields suffers no company would be worse off after sharing in these benefits.

6.5 THE OUTCOME OF TAX DELIBERATIONS

The 1982 Budget Changes

Following the discussion and deliberations after the Chancellor invited proposals for tax reform, the tax changes that were implemented represented no more than a tinkering within an unchanged tax system.[14] In essence the Chancellor opted for the tax structure proposed by UKOOA, but with rates and allowances set at levels to generate the government's requirements for tax revenue. UKOOA's replacing of SPD by APRT was accepted, but the gradual withdrawal of APRT was ignored. The additional oil allowance in PRT suggested by UKOOA was ignored also, while the PRT rate was increased to make good the likely shortfall in tax revenues caused by APRT replacing SPD. In accepting some features suggested by the industry, the government appeared to believe it had solved the problems of marginal field developments since the industry had a tax structure which was not too dissimilar from its stated needs. Although, of course, the industry would have preferred lower tax rates and greater allowances, the government appeared to believe that the distribution of the tax burden across fields was acceptable and that the industry would not object to the burden on marginal fields.

The 1983 Budget Changes

The Chancellor introduced in the March 1983 Budget three tax changes which significantly benefited expected profitability of future fields. Only one of the changes applied to projects already under way (those with Annex Bs granted prior to 1 April 1982), while all three applied to future developments. The change applied to existing fields in the phased abolition of APRT by staged reductions in its rate up to 1987. Fields granted development approval after 1 April 1982 benefit not only from this measure but also from the abolition of royalty payments and the doubling of the PRT oil allowance (both per period and per field). There were also changes to the tax treatment of exploration expenditures which operate to the private sector's advantage.[15]

It is worth emphasising that it is very difficult, and perhaps impossible, to help marginal fields without reducing aggregate tax revenues whilst keeping the original structure of taxation. This point appears to have been borne out by the 1983 Budget tax changes which attempt to assist new marginal fields without reducing aggregate tax revenues significantly but which do not stay within the original tax structure. By distinguishing between new and old fields the taxation of North Sea oil had a new dimension that took it outside the original structure. Thus in order not to distort significantly tax revenue in the immediate future the tax system applying to existing discoveries was not altered substantially. At the same time, to stimulate more development activity, new fields faced a considerably more lenient tax structure.

The effect of abolishing APRT – the only change that affects short-term aggregate tax revenues – is only a slight reduction in government receipts. As the bulk of PRT over the next few years is likely to be paid in mainstream PRT rather than in APRT, and as the reduction in the rate of APRT is in gradual stages, the effect on short-term tax revenues is minimal. The official government estimate[16] was that tax revenues between 1983 and 1987 will be less than $2\frac{1}{2}$ per cent lower because of this change. The government's criteria for North Sea taxation seem to be satisfied.

Estimates of the boost to field-by-field rates of return on new discoveries suggest the industry would welcome the tax changes. Typical real rates of return are expected to improve by as much as 4 to 6 percentage points, as a result of the 1983 tax changes, and real NPVs by about £1 to £2 per tonne. All fields show significantly increased profits, and the decision to submit development programmes was taken for several projects (for instance, Duncan, Balmoral and Highlander).

The industry welcomed the tax changes and the government seemed content as its short-term tax revenues were not likely to decline markedly. It thus appears that the problems of North Sea taxation were resolved by changes announced in the 1983 Budget.

The 1984 Budget Changes

Reductions in the rate of CT, announced in the 1984 Budget, should have been welcomed by North Sea producers as they lessened the distortionary impact of CT. However, the reduced first-year expected allowance for CT had a detrimental impact on companies involved with multi-field projects. As with the 1983 tax changes, the 1984 changes, on balance, leave government short-term tax revenues by and large unchanged.

The Structure and Dynamics of Taxation: Problems Resolved?

New field profitability is higher under the current tax system than before the 1981/2 period of tax deliberations, but more importantly the structure of taxation has improved somewhat. The major tax changes since 1982 include increasing the PRT rate, the abolition of SPD and phasing-out of APRT, a reduction in the CT rate, a reduction in the CT capital allowance, an increase in the PRT oil allowance and the abolition of royalties on new fields. Although many of the changes will have improved profitability, only two of the changes (the abolition of royalty payments on new fields and the abolition of SPD and phasing out of APRT) are likely to have improved the tax structure. On the other hand, two changes (reduction in CT capital allowance and the increase in the PRT oil allowance) are likely to have harmed the structure.

Chapter 3 emphasised how regressive elements of taxation deter development decisions and thus abolishing tax features which are not related to costs – such as royalties and APRT – can only be welcomed.

In particular, the abolition of royalties must be strongly applauded as significantly improving the North Sea tax structure. Generally, the greatest benefits from abolishing royalties are enjoyed by the less profitable potentially commercial fields, in particular Claymore North, Highlander and Clyde (the latter pair have since been granted development approval), all of whose NPVs were increased by over £1 per tonne.

Although APRT is also a regressive feature, the damage to development decisions did not seem to be on the same scale as the damage caused by royalties. Accordingly the improvement to field profitability from abolishing APRT is smaller than the improvement from abolishing royalties. Moreover, abolishing APRT gives a roughly even spread of benefits across the potentially commercial fields.

Conversely, the decision to double the PRT oil allowance is not supported by our analysis as it is likely to bring a deterioration in the inter-field burden of taxation. The oil allowance's benefits were found to be proportionately greater for the fields least in need of any benefits, and hence increasing the allowance is not likely to improve the structure of North Sea taxation. The benefits of increasing the allowance are related to the size of field reserves, and to a lesser extent to the shape of the production profile, which are poor indicators of field profitability and in practice tend to accrue to the more profitable fields (for example, Thistle NE, Columba, Mabel and Andrew); that is, the fields which would be likely to be developed in the absence of this tax change anyway. Hence, this tax change is unlikely to improve marginal prospects and is unlikely to increase recoverable reserves more than negligibly.

The main adverse impact of the 1984 tax changes is on the return to capital for companies with multi-field North Sea interests. Capital costs are deducted from CT profits in historic terms with no allowance for inflation or the normal real rate of return on capital. The time lag between incurring capital costs and the reduction in CT payments, however, will have lengthened as a result of the reduction of the first-year capital allowance from 100 per cent to 25 per cent and the detrimental impact of CT on the normal return to capital will – *ceteris paribus* – have been aggravated. Nevertheless, the reduced CT rates will have reduced the overall burden of CT.

Overall the structure of taxation does seem to have been improved by the sequence of tax changes. Since 1982 with the current tax system, the reduction in NPVs as a percentage of the pre-tax NPV is only greater than 100 per cent in three instances (Clyde, Clair and

Alwyn South). However, this measure of the inter-field impact of taxation suggests that the tax burden is still badly distributed. Tax accounts for less than 80 per cent of the pre-tax NPV on all except one of the most attractive eight potentially commercial fields, while accounting for more than 80 per cent on five of the six least attractive fields.

Chapter 3 drew attention to two major faults in the North Sea tax structure: first, it fails to operate progressively and second, it acts as a tax on capital. Our analysis indicated that the second of these faults has a greater detrimental impact on development decision. No attempt at solving the second fault has been made and hence it is not surprising that problems of taxation persist.

Analysing the tax changes in the light of the dynamics of taxation suggests a further, rather sombre, conclusion. The recent tax changes were completely in line with our earlier reasoning which identified the causal link between dollar oil prices and the harshness of the tax system, with the latest reductions in dollar oil prices causing the government to respond by reducing its tax take on new projects. This leads to a sombre conclusion because it reinforces the argument that if (or perhaps, when) dollar oil prices start to rise the tax changes will be reversed. Not only would this reversal undermine any stability in the UK oil industry, but the scale of the reversal may very well more than offset the reduction in the level of tax made since 1982.

6.6 A RECURRING ISSUE

Despite acceptance of the current tax system, the underlying problems have not been resolved. The prevailing contentment with North Sea taxation suggests the problems of marginal field taxation will tend to be ignored. This would be unfortunate: numerous new discoveries would still fail to be developed even though their development could make both the government and the industry better off. We doubt, though, if the prevailing contentment will last. As soon as oil prices start to rise then government and industry perceptions of their rewards from the North Sea may tend to diverge. Marginal field problems, however, are not likely to be of paramount importance when oil prices are rising and when the government is likely to be looking to increase the level of taxation, so the contentment seems unlikely to last. The scale of the damage caused by the inter-field burden is equally unlikely to be attenuated.

Finding a Solution

Whenever the tax system is debated, it appears that the industry will attempt to lower the total tax take by seeking adjustments to the level of taxation. Government, however, seem at most prepared to let the industry determine the structure of taxation, setting the level of taxation itself. Although tax changes may follow debate on the tax system between government and industry, the tax position facing marginal fields is unlikely to be improved and investment is still likely to be deterred.

Several possible events could bring the tax system into debate. There seems to be a connection between oil price changes and tax changes, but this connection is only one, albeit the most important, of many forces which direct politicians' attention to oil tax policy. Whenever politicians perceive a change that benefits the industry, they are likely to want to participate in the benefits. Apart from a change in oil prices, an instigating event could be the discovery of a rich oil–bearing basin, some technological advance that brings about a drastic decline in oil–field costs, or anything that makes the oil sector easier to tax without losing votes, which will occur whenever the electorate perceives an increase in the capacity of the industry to bear taxes. One factor which seems unlikely to bring the tax system into debate, however, is concern for marginal field taxation. Perhaps concern over the lack of orders in the offshore supply industry will increase, but until the extent of damage caused by taxation is established, concern for marginal fields is unlikely to be strong enough to motivate either private or public sector agents. Although tax changes may not be infrequent, the debate is unlikely to be fruitful for marginal fields and the perilous equilibrium is likely to persist.

Implementing a Solution

One solution to the taxation problem is given in the IFS proposals. This is not the only solution, but for any change to be implemented two inter-related requirements are necessary. First, implementation would occur if the IFS proposals were viewed by all as an improvement. To establish this improvement, the potential benefits from the tax changes would need to be demonstrated to all concerned. This is an onerous task. Not only is the extent of damage not appreciated, but it can never be proved. Second, the industry would need to drop any aspirations to lowering the total level of taxation. Both these requirements involve perceptions about 'what might otherwise happen'.

Having established the extent of the damage and having established the constraint of government tax requirements, we hope that this book will help to change those perceptions.

Onus on Government

We have suggested that it is naive of the government to rely on the industry for suggestions on ways to improve the structure of taxation. The industry will always tend to base suggestions on lowering the total level of taxation. These suggestions, though, as we have seen, seem to be based on a false perception of government tax requirements. In the absence of any change in these perceptions, the onus must be on the government to improve marginal incentives. Improvements are available which the government should coerce the industry into accepting. For the government to be willing to implement PPT it must believe it will be better off. We have concentrated on the consequences for aggregate tax revenues in order to demonstrate the impact on government, while in this context the inter-field tax burden need not be explored in great detail. This revealed the direct impact on government. The government would also be concerned about any indirect impacts. In particular, if some companies feel worse off because of the changes to PPT, their support for the government may dwindle. To this end we have provided evidence that shows the boost to North Sea supplies, and hence the gains enjoyed by the private sector, would be substantial if PPT were adopted. We believe the sheer scale of the problem is not commonly appreciated. If if were to be appreciated then we doubt if there would be many companies who would feel worse off, unless they still hold aspirations to reduce the level of taxation.

If the government can be persuaded of its role in determining the structure of taxation, instead of relying on industry recommendations, we will come closer to achieving a solution. The first step in altering politicians' views is to establish the detrimental consequences of ignoring the problem. Ignoring the problem may be politically expedient – since suppressing difficult areas of policy is a safe option – but ignoring the detrimental consequences does not alleviate them.

Appendix A: Description of the North Sea Fiscal Regime

Essentially the current system of North Sea taxation comprises two separate systems, one for Southern Basin fields and fields granted development approval before April 1982, and one for fields (except Southern Basin fields) granted development consent on or after 1 April 1982. Since our interest is the impact of taxation on new development projects we are primarily concerned with the latter system, which is referred to as the system applying to 'new' fields.

Whereas for old fields a three-tier system applies comprising royalties, CT and PRT, for new fields a two-tier system operates involving only CT and PRT. PRT is a six-monthly charge on each oil field and is a tax on the landed value of gross revenues less several deductions, and for a defined period of time is subject to a ceiling limit. This calculation determines the mainstream liability to PRT. There is also an assessment of APRT, which shifts the timing of tax liability to earlier periods but does not alter total lifetime liability, and regulations for the payment of liabilities to spread them throughout each six-monthly chargeable period. With respect to all fields, APRT is being phased out and as from 1 January 1987 will be abolished. APRT thus has no impact on new development projects.

In calculating PRT profits, all operating expenditures and all capital expenditures can be deducted immediately from the landed value of gross revenues. Moreover, capital expenditures before the date when PRT profits are first earned attract an 'uplift' which can also be deducted immediately. The uplift is equal to 35 per cent of capital expenditure so that 135 per cent of capital costs can be deducted. After PRT profits have been earned (the time of PRT payback), the uplift is no longer available and only 100 per cent of subsequent capital expenditure may be deducted. When these deductions exceed gross revenues (landed value), the excess is carried forward and deducted in later periods.

Once PRT profits are positive an oil allowance may be used to reduce the profits. The oil allowance exempts the first 0.5 million tonnes of oil per six-monthly period from PRT, up to a total limit of 10 million tonnes of oil per field in the 'new' system. For older fields the six-monthly allowance is 0.25 million tonnes up to a field limit of 5 million tonnes. Profits remaining after this deduction are charged at a rate of 75 per cent to yield each period's untapered (see below) liability.

A tapering limit may act as a ceiling to reduce the PRT liability. This limit applies for a period after PRT payback equal to half the length of time between production start-up and PRT payback. The limit is calculated every six months and is 80 per cent of gross revenues (landed value) less two deductions. The deductions are operating costs and 15 per cent of accumulated capital costs (also royalties under the old system). If the untapered PRT liability exceeds this calculation, PRT is limited to this amount; otherwise the limit is ignored.

PRT payments are settled two months after the end of each six-monthly period. Pre-payment of six equal instalments is made in each of the previous six months. The pre-payments amount to 75 per cent of the previous period's liability. Hence, at the time of settlement, the payment will equal this period's liability less 75 per cent of the previous period's liability. In any six-month period, payment will equal this amount plus two instalments of pre-payments for this period's liability plus four instalments of pre-payments on next period's liability. As from January 1983 the rate of PRT has been 75 per cent.

The second tier of the 'new' system is CT. CT is charged on the profits of oil companies operating in the North Sea. There is a ring fence to prohibit the deduction of losses made outside the North Sea from North Sea profits, but otherwise the CT treatment of offshore oil profits is identical to the treatment of other profits. CT profits are gross revenues (landed value) less a deduction for capital costs, less all operating expenses, less interest on debt, less a deduction for PRT payments and less a deduction for capital costs based on a first-year allowance undergoing phased reduction to 25 per cent by April 1986 and a 25 per cent declining balance method of depreciation thereafter. CT losses may be carried forward and deducted from profits in subsequent years. The rate of CT was reduced from 52 per cent to 50 per cent for the 1983–4 financial year in the 1984 Budget, and is scheduled to drop to 45 per cent for 1984–5, 40 per cent for 1985–6 and 35 per cent thereafter.

Numerous changes have been made during the evolution of North Sea profits taxation and it should be emphasised that the system outlined above applies only to 'new' projects (not in the Southern Basin). The evolution of the North Sea fiscal regime is shown in Table A.1 below. The chart concentrates on the main features of the tax system and illustrates the system's instability over time, a theme which is examined in Chapter 3.

Table 4.1 The evolution of the North Sea tax system

Date of proposal	*Date of implementation*	*Description of change*
March 1975	January 1975	PRT introduced at a 45% rate with 175% uplift, annual oil allowance of 1m tonnes, field oil allowance of 10m tonnes and tapering provision.
August 1978	January 1979	PRT rate increased to 60%.
	July 1978	Uplift decreased to 135%.
	January 1979	Oil allowance in PRT halved. Payment of royalties accelerated.
March 1980	January 1980	PRT rate increased to 70%.
	January 1981	Advanced payments of PRT introduced.
March 1981	January 1981	SPD introduced.
	January 1982	Uplift limited to period up to PRT payback. Tapering and safeguard limited to a period half as long as the period to PRT payback, once PRT payback is reached.
March 1982	January 1983	SPD abolished, APRT introduced, PRT rate increased to 75%.
	July 1983	Advanced payments of PRT replaced by a system of spreading PRT payments.
June 1982	January 1983	APRT applies for 5 years only with any outstanding APRT settled at the end of the fifth year.
March 1983	July 1983	APRT reduced to 15%.
	January 1985	APRT reduced to 10%.
	January 1986	APRT reduced to 5%.
	January 1987	APRT abolished. All exploration and appraisal expenditure can be deducted from PRT profits on any field. For fields granted development approval after April 1982, excluding Southern Basin fields:

Table 4.1 The evolution of the North Sea tax system

Date of proposal	*Date of implementation*	*Description of change*
		(a) PRT oil allowance doubled to 10m tonnes per field and 1m tonnes per year per field; (b) royalty payments abolished.
March 1984	April 1983	CT rate reduced to 50%.
	April 1984	CT rate reduced to 45%.
	April 1985	CT rate reduced to 40%.
	April 1986	CT rate reduced to 35%.
	April 1984	CT first-year capital allowance reduced to 75%.
	April 1985	CT first-year capital allowance reduced to 50%.
	April 1986	CT first-year capital allowance reduced to 25%.

Appendix B: How Much Oil is There? An Econometric Analysis of Oil Reserves on the UKCS

1 INTRODUCTION

In this appendix a behavioural model of the oil extraction industry is constructed. This model permits UK offshore recoverable oil reserves to be quantified under a variety of economic and geological assumptions, and under a variety of hypothetical tax structures. Chapter 4 employs the model to estimate the effect of current tax policy on the extent of recoverable reserves.

Other Studies

Non-renewable energy supply is an ambiguous title for two inter-related but different topics. First, it relates to the timing of production – whether to produce now or later – and second, to the total volume of reserves that will be extracted from a resource bearing deposit during its life. The first topic has received much attention from economists in both applied and academic research,[1] where the incentives facing resource explorers and extractors have been identified, quantified and categorised in order to pinpoint the scale and direction of a desirable depletion policy. Although no unquestionable conclusion can yet be drawn from the research, the arguments are well known.

Paradoxically, the more fundamental topic concerning the size of the non–renewable resource base has received much less attention. Estimates of reserves are made according to *ad hoc* economic, geological or administrative principles which can be only vaguely interpreted by anyone other than the estimator and which tend to have little economic meaning. The commonest classification of reserves splits an oil basin into proven, probable and possible reserves:[2] that is proven, probable and possible for given (but usually unstated) economic conditions. While this classification has been meaningful in the past, since the early 1970s extraction costs and especially oil prices have become so volatile that the classification is no longer particularly helpful.

Instead, the economic conditions which determine oil supply must be considered explicitly rather than implicitly. Only a few attempts have been made along these lines, none of which seem appropriate for our analysis of the relationship between tax policy and UKCS oil reserves south of the 62° latitude. Hence we have developed our own methodology, explained in the next section.

2 METHODOLOGY: GENERAL

The Nature of Oil Deposit

If we wish to estimate recoverable oil reserves we must first recognise the nature of an oil deposit. Oil accumulates by moving through permeable rocks until it reaches a position where its driving force is insufficient to move it further.[3] Where the permeable oil-bearing rocks are sealed by impermeable surfaces then the density of oil will be greater and the deposit will be easier to extract. As most rocks hold some oil, it is clear that calculations of reserves-in-place will yield massive but meaningless figure. Instead we need to know how easy it is to extract each barrel of oil from the deposit, or equivalently, how costly. To determine how costly the extraction of each barrel is, we need to know who is extracting the oil and on what criteria their investment decisions are based.

The Corporate Role

In the UK, private companies seeking to maximise profits as one of several operational criteria are by and large responsible for extracting oil, and thus we need to assess corporate criteria. For each oil prospect, if the companies comprising a consortium consider all or part of an oil accumulation to be worthwhile then this part of the oil underground should be counted as recoverable reserves; if the companies in the consortium do not consider the prospect to be viable then the oil underground should not enter our reserve estimate. The factors determining any company's willingness to extract oil from a deposit are varied and numerous. Among the most important factors are:

(a) the geological, geographic and climatic circumstances of the prospect, which determine the expected development and operating costs;
(b) the expected price of oil, which with production rates determines the likely revenue profile;
(c) government taxation or depletion control policies towards oil producers; and
(d) the available investment options which the company might otherwise pursue if it did not invest in the UK oil prospect.

If all these factors were known then it would be a trivial exercise to quantify UK reserves.

The Influence of Prices

Perhaps the most crucial (but certainly the most volatile) of these factors is the price of crude oil. Unfortunately, the influence of prices on reserve estimates is not explicitly recognised in official publications. For example, the Department of Energy's annual Brown Book[4] gives a range for reserve estimates without commenting on the underlying oil price assumptions, and it is not even clear whether the implicit prices assumed in reserve calculations are consistent with official assumptions made elsewhere. Without a notion of price the reserve estimates cannot be interpreted.

Outside official circles there are few published studies which recognise the influence of prices.[5] Without exception these studies determine the minimum field size which is commercially viable at a given price, and with a distribution of field sizes in place, use this to devise reserve estimates for the UKCS. As large fields are assumed to have lower unit costs, with low oil prices large fields alone are assumed to be exploited. As prices rise, smaller fields become worthwhile and the commercial reserve base expands.

Naturally much attention is paid to the distribution of field sizes in place. Indeed some models[6] are exclusively concerned with this distribution and use estimates of this distribution alone to estimate recoverable reserves. None of these models uses econometric techniques to estimate the parameters and consequently they all tend to pass on the modeller's subjective interpretation of geological data.

Apart from our unease over methods which rely on purely subjective interpretation of geological data, there are two fundamental reasons why we cannot adopt the 'minimum commercial field size' approach. First, experience to date suggests that larger fields do not necessarily have lower unit costs. The theory cannot account for two of the earliest North Sea discoveries – Argyll and Auk – which are both small but commercially attractive under even pessimistic price forecasts. Nor can the theory explain the large unit cost differentials between fields of roughly the same size (and in some instances of neighbouring localities): for example, Cormorant North, Fulmar, Magnus and Thistle are each believed to hold around 500 million barrels (70 million tonnes) of recoverable oil yet their costs including a return on capital[7] range from under £40 per tonne to about £70 per tonne (in real 1983 sterling terms). North Sea experience is not consistent with the theory that size is solely related to profitability. Second, even if large fields are more profitable, the approach presupposes that the impact of the tax system is such that more profitable fields pay proportionately more tax. Most commentators deny this proposition.[8] Moreover, we wish to use our model to asses this proposition and hence need to develop our model independently of such assertions.

Cumulative Approach

Although we dismiss the 'minimum commercial fields size' approach, we do utilise some similar concepts, notably that of 'supply price'. But rather than concentrating on how the characteristics of marginally commercial fields change as oil prices change, our methodology constructs a cumulative supply curve showing how total recoverable reserves expand as oil prices increase.

Considering only extensively explored regions, we impute from expected cost data and tax schedules the price at which each and every field in a region becomes commercially viable. At different oil prices we can infer the level of recoverable reserves from the region by cumulating only those fields which would be commercially viable. This yields a scatter of points on a simulated supply curve: points which are then used to estimate the true supply curve.

3 METHODOLOGY: SPECIFIC

Data Area

Under the first four licence rounds (completed in 1964, 1965, 1970 and 1972) some 863 blocks were issued that have, by now, been fairly extensively explored. The licensing regulations ensure that at least half of the licensed acreage is relinquished after six years encouraging operators to explore the licensed area in order that the better portion is not relinquished. In fact, closer to three-quarters of the acreage has been relinquished.[9] New discoveries on this acreage are unlikely. For the fifth (completed in 1977) and subsequent licence rounds exploration is not yet complete and further discoveries are likely. We must then restrict our attention to the acreage licenced prior to 1976, when it is reasonable to presume that, by and large, all the finds that will be made have been made. A find such as Clyde, made on a fifth round licence, is not relevant to our study.

Not all the 863 blocks[10] cover potential oil-bearing strata. A large number cover potentially gas-bearing reservoirs off the coast of East Anglia and some cover areas that are not over recognised geological basins. Three oil-bearing basins, however, were covered quite extensively – the East Shetland basin, the northern North Sea Basin and the Moray Firth basin[11] – and it is these licensed blocks (totalling some 238 blocks) to which our database must be restricted. Hence we will exclude from consideration discoveries such as Clair, located to the west of the Shetlands.

Particulars of the finds made on the database blocks vary widely. Indeed, many are too complex, too remote and too thinly formed ever to be of any commercial interest. There are, however, a number of commercial prospects amongst these finds. Some 30 of the 31 'established commercial' oil fields that have been declared to be viable by their operators are finds made on licences issued prior to 1976 and are located in one of the three basins listed above. Also, there are also 24 fields currently undergoing further appraisal.

Simulated Database

As our basic source of information is declared company plans, we are only likely to know of the fields which are viable at current oil prices and oil price

Figure B.1

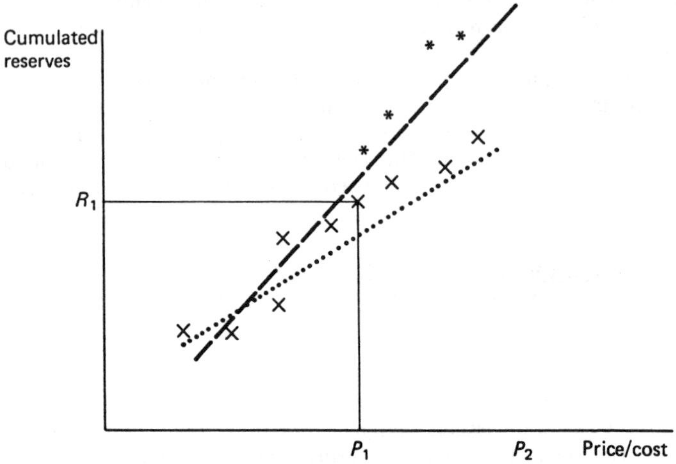

X = Known data points
* = Unknown data points
--- = Properly estimated line
.... = Improperly estimated line

expectations. We may know of an occasional discovery which would be commercial at prices higher than currently expected, but we would be unlikely to know of *all* the projects viable at this higher price. So while we may be justified in supposing all finds viable at current prices have been announced, we would not be justified in supposing that the announced fields include all eventual discoveries. Indeed, this is one purpose of the estimations: to forecast the extent of additional discoveries at higher oil prices. It follows that we must devise some means to select from the many announced finds those which are (or will be) viable at current prices.

Figure B.1 illustrates the type of bias that might enter our analysis if we do not tread carefully. If we rank all the discoveries by their real unit costs (including a normal return on capital and a cost of bearing risks) and then cumulate recoverable reserves at each cost, we might arrive at the crosses shown in the diagram. So long as the costs include all outgoings then with a real oil price of P_1, R_1 reserves would be commercially recoverable. Assuming current real oil prices have been established for some time at P_1, then it is likely that no unannounced viable finds (given a real oil price of P_1) exist in this well-explored acreage. At P_2, however, there may be many unannounced finds, whose inclusion would lead to the points marked with an asterisk. Estimating the supply curve on all the crosses (the dotted line) would be misleading. Instead only the points to the left of P_1 should be included (yielding the dashed line).

The Representative Firm

Selecting the fields to be included in the database is not straightforward. For each field we have found the 'supply oil price' which just makes it worthwhile and just outweighs the risks, the costs, the disadvantages of tied-up capital and the taxes. Comparing these supply oil prices with actual oil prices will indicate whether the field should be included in or excluded from the database. However, the supply oil price for any one project is not unique. An oil project's commercial valuation, as determined by the consortium exploiting it, will depend on a variety of factors external to the project itself; each company's alternative investment opportunities, each company's expectations of inflation, exchange rates and oil price growth, and each company's corporate tax position will affect a project's commercial attractiveness. For example, even if future development and operating costs could be accurately forecast, even if expectations on oil price growth were identical between companies, and even if the companies faced identical choices, if some companies could use the project capital costs to delay other CT liabilities then that project's development would seem more attractive than when viewed by companies without other CT liabilities. To overcome these differences and produce a unique 'supply oil price' for a given project, we adopt the notion of a representative firm which has one outlook for oil price growth, one outlook for sterling values, one idea on cost inflation, one alternative opportunity and one CT position. Clearly, establishing the sensitivity of our results to assumptions about our representative firm is an important exercise.

Our representative firm is assumed to have no access to loan finance; it is interested in one discovery only and has no other assets generating a CT liability; it expects oil prices to stay constant in real terms from whatever contemporary level is reached; it expects capital cost inflation to drop to 5 per cent per year and the exchange rate to remain fixed at £1 = $1.45; it has an alternative risk-free project which would yield a real 5 per cent per annum return that would be displaced if the oil discovery is developed; and it requires a risk premium equal to the discounted value of each field's capital costs before being willing to bear the risks attached to operating a North Sea project. If our representative firm is not truly representative then some bias will be imparted to our estimates.

Excluded Data

Some peculiarities arise from our definition of the representative firm. For example, our calculations (outlined below) show over $30 per barrel supply oil prices for Heather: that is, the established oil price would have to be some $32 per barrel (in real 1983 terms) before our representative firm would consider this field to be worthwhile. However, even with long–run oil price expectations of $25 per barrel the operator is in the process of developing this discovery. There appear to be two explanations for this peculiarity. First, the operators can use the capital costs on these fields to delay CT liabilities on other North Sea or onshore profit-making projects. This enhances the attractiveness of some fields. Second, costs of several fields have risen much faster than general cost inflation on North Sea projects. BP's conversion of the semi-submersible rig for use as Buchan's producing platform has involved unparalleled cost overruns and is an experiment which seems unlikely to be repeated. Cormorant's costs are artificially high because much of Shell/Esso's North Sea pipeline network utilises separators and compressors on the Cormorant platform, and consequently is costed to the Cormorant project although it is used by many fields. Heather's operating costs are proving to be several times as great as originally expected. Hence we suggest that if the operators of Buchan and Heather were to reappraise the economics of these fields with today's knowledge, they might be reluctant to develop them. Similarly, if Cormorant had to prove its viability on its own merits, without considering the services it provides for other fields, then it might not be pursued. Thus, although these three fields are commercial and are well under way, our evaluation procedure indicates that they must be excluded from our database since at current prices our representative firm would not develop them. If we were to include these discoveries our model would underestimate cumulative reserves as demonstrated in Figure B.1 and discussed in section 3.2.

Established Oil Prices and Estimation Technique

The problem of excluding data is compounded because the current established price of oil is not, and never can be, uniquely defined. The 'established' price of oil in the sense used here relates to a long-run price and corresponds to an average of the expected real prices to be received over the life of the field. It is based on corporate expectations and thus will be influenced by contemporary real oil prices. If expectations are formed rapidly and are changed as soon as oil prices change, then the established price will closely mirror contemporary prices. These conditions, however, are rather unlikely to characterise oil companies. Instead, the established oil price will probably move sluggishly, corresponding to a moving average over the past few years of real oil prices. To nullify the bias from this difficulty in the following regression analysis, we place more emphasis on finds with lower supply prices than on our data points at higher oil prices.

We are justified in placing this emphasis because the trend over the 15-year period to 1980 has seen oil prices move upwards and hence the data points at higher oil prices are associated with more recent times. Operationally, this emphasis is implemented by choosing the weighted least squares estimation procedure.

The Data

Using our representative firm to simulate development decisions shows that the area licensed in rounds one to four for the three potentially oil-bearing basins holds some 54 commercial discoveries. Our database must rely on only these fields: 29 in the East Shetland basin, 14 in the northern North Sea basin and 11 in the Moray Firth basin. For each of these discoveries we have compiled year-by-year estimates of likely operating costs, development costs and oil output, drawn from the most reliable and up-to-date sources available, from discussions with oil companies and from the trade press. On the basis of this data we have calculated over the life of each field:

(a) the oil price that would be just sufficient to outweigh the costs, including a normal return on capital and a risk premium but excluding taxes and royalties, and just sufficient to induce our representative company to proceed with development: which we term the 'economic supply oil price'; and
(b) the oil price just sufficient to outweigh the costs plus any royalties, PRT, SPD or CT that would be due at that price under a given tax system and just sufficient to induce our representative company to consider the field a commercial proposition: which we term the 'commercial supply oil price'.

Throughout this appendix we use the term 'economic' to refer to a prospect where revenues exceed costs (including a normal return to capital and risk premium) while the term 'commercial' implies revenues exceed costs (including a normal return to capital and risk premium) plus taxes. This simulated and computed data is categorised in Table B.1 by geological basin and is the crux of our analysis.

4 ESTIMATING RECOVERABLE RESERVES

Asymptotic Reserves

For each of the three basins we cumulate the reserves that are economically recoverable at various oil prices to yield three supply schedules as in Figures B.2a, b. For the moment we ignore taxation. The schedule shows how in each basin, as oil prices rise, more fields' revenues exceed costs so that economic reserves increase. Economic reserves cannot continue rising forever, and at some point further price rises will no longer bring forth additional supplies. At this point, all the reserves in place will be included in economic reserves and we will have the asymptotic reserve estimate. Although this estimate is of little importance in itself, it must be quantified before we can proceed.

Using a single equation model we can attempt to identify the relationship between total recoverable reserves and oil prices for each of the three basins. On *a priori* grounds as explained in section 2 above, the relationship must be symptotic. After testing several forms, the most suitable structure for its simplicity and for its statistical and econometric properties[12] is as follows:

$$R = La\, e^{-b/p}$$

where R represents recoverable reserves from the licensed part of a basin, L

Table B.1 Prices giving a real 5% return plus risk coverage equal to discounted field capital costs

	Reserves (million tonnes)	*Economic supply oil price (real 1983 $/bbl)*	*Cumulated reserves (million tonnes)*	*Commercial supply oil price (real 1983 $/bbl)*	*Cumulated reserves (million tonnes)*
East Shetland Basin					
Alwyn North	26	17.6	1244	20.5	1244
Alwyn South	17	16.3	1144	18.0	1054
Beryl	94	12.1	803	13.7	826
Brae	43	10.8	631	12.4	293
Brae North	26	9.9	166	10.9	166
Brent	265	10.3	515	13.3	631
Bruce	22	22.6	1284	25.4	1284
Columba	24	12.4	826	13.4	655
Cormorant	5	18.9	1249	22.6	1262
Cormorant North	78	11.3	709	13.8	732
Crawford	44	6.3	46	7.2	46
Deveron	2	2.3	2	2.5	2
Don	12	19.7	1262	21.5	1256
Dunlin	41	12.8	867	17.8	1029
Eider	17	16.9	1199	18.8	1121
Emerald	8	16.0	1127	18.0	1037
Heather	10	25.9	1293	32.3	1293
Hutton	31	15.8	1119	18.6	1104
Hutton NW	19	17.1	1218	19.4	1198
Lyell	36	9.9	203	11.1	203
Magnus	73	10.6	588	12.4	366
Maureen	20	16.5	1164	20.2	1218
Murchison UK	40	7.8	86	10.8	140
Ninian	140	13.3	1008	16.8	967
Statfjord UK	54	7.9	140	9.4	100
T-Block	47	10.2	250	11.4	250
Tern	21	15.3	1088	17.0	988
Thistle	59	13.7	1066	19.4	1179
003/28	18	16.7	1182	18.2	1072
Northern North Sea Basin					
Andrew	14	16.9	387	18.6	387
Argyll	9	18.9	402	22.3	408
Auk	8	21.8	417	28.3	417
Balmoral	13	12.1	333	13.5	333
Duncan	3	15.4	364	17.7	364
Forties	266	6.1	266	10.3	321
Fulmar	54	7.9	321	9.7	54

Table B.1 Prices giving a real 5% return plus risk coverage equal to discounted field capital costs

	Reserves (million tonnes)	*Economic supply oil price (real 1983 $/bbl)*	*Cumulated reserves (million tonnes)*	*Commercial supply oil price (real 1983 $/bb1)*	*Cumulated reserves (million tonnes)*
Glamis	3	13.2	347	15.4	347
Kittiwake	9	15.8	373	17.9	373
Mabel	11	12.8	344	14.8	344
Montrose	12	23.9	428	28.8	428
Montrose South	6	18.7	393	20.7	393
030/7	15	13.9	361	15.6	361
030/13	6	20.0	408	22.1	399
Moray Firth Basin					
Beatrice	19	19.0	257	22.7	257
Buchan	11	19.1	268	25.2	268
Claymore	52	10.2	179	13.6	217
Claymore North	6	18.3	239	20.7	239
Ettrick	14	11.4	217	13.1	165
Galley	24	10.3	203	12.0	151
Highlander	7	16.8	233	18.6	233
Piper	109	5.7	109	8.4	109
Renee	18	9.7	127	10.6	127
Scapa	9	13.3	226	14.5	226
Tartan	11	25.4	279	28.8	279

is the number of licensed blocks in the basin, reflecting the size of the licensed area, *a* represents asymptotic reserves (that is, reserves in place) per average block in that basin – to be estimated, and *b* represents growth parameters reflecting the speed with which the asymptote is reached – to be estimated.

The shape of this logistic curve, shown in Figure B.3, gives a good approximation of the basic geological realities. The curve rises initially only very slowly reflecting the existence of significant fixed costs, then rises at a fairly uniform rate reflecting the ease of extraction in that basin (which is unlikely to vary significantly for any given basin since the water depths, rock conditions and climatic factors are generally common), and then rises towards an upper asymptote which may be thought of as reserves in place.

As we wish to find reserves in place, our interest centres on the ease of extraction from the oil-bearing rocks as quantified by the resources (costs plus normal return and risk premium) used to develop a find, that is the

Figure B.2a Simulated data points compared to model estimates in the East Shetland basin

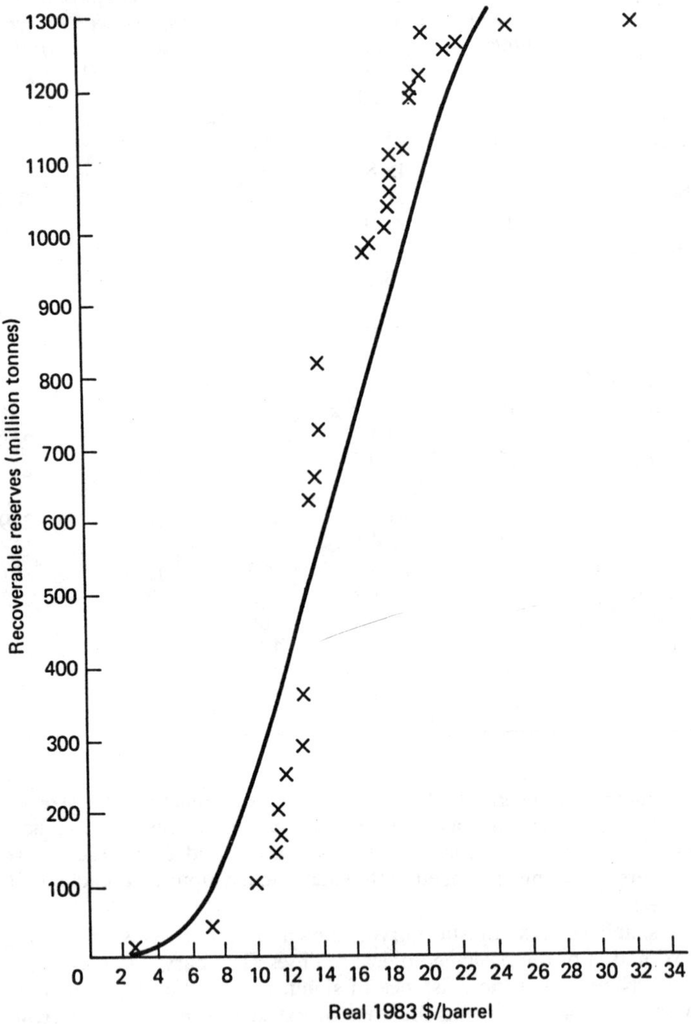

XX = Data point
——— = Model estimates

Figure B.2b Simulated data points compared to model estimates in the Moray Firth basin

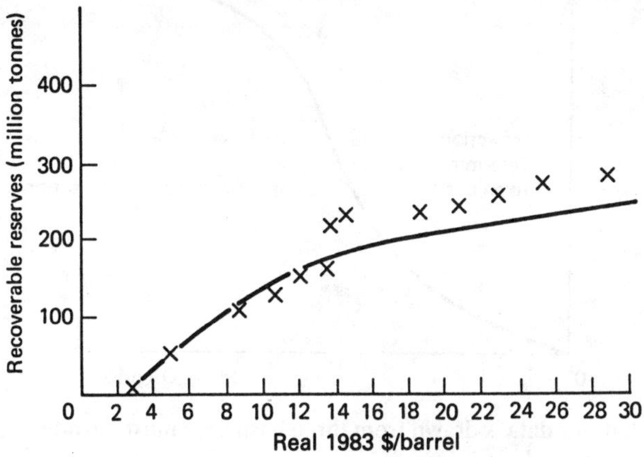

Figure B.2c Simulated data points compared to model estimates in the Northern North Sea basin

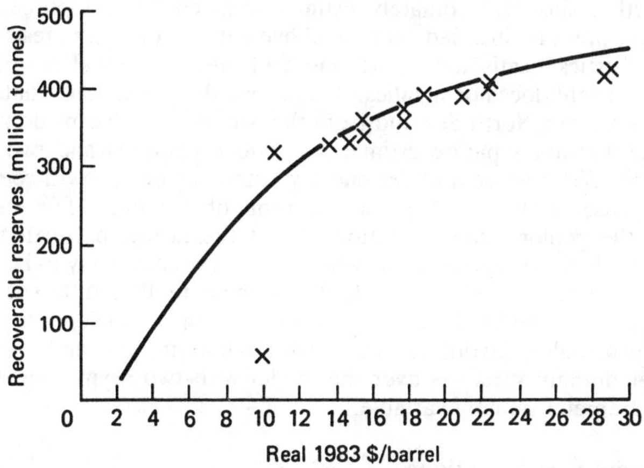

economic supply oil price (*Pe*). Costs to the company which merely reflect a transfer of resources from the company to other groups, such as taxation, should be ignored. The geology and the size of each basin will, by and large, determine the asymptote '*a*': the basin size is relevant as smaller basins will probably not have the same pressure to drive and compact oil. The growth parameter '*b*' will reflect the hostility of each area's climate and the local water depths, with a larger 'b' indicating a more hostile environment.

Figure B.3

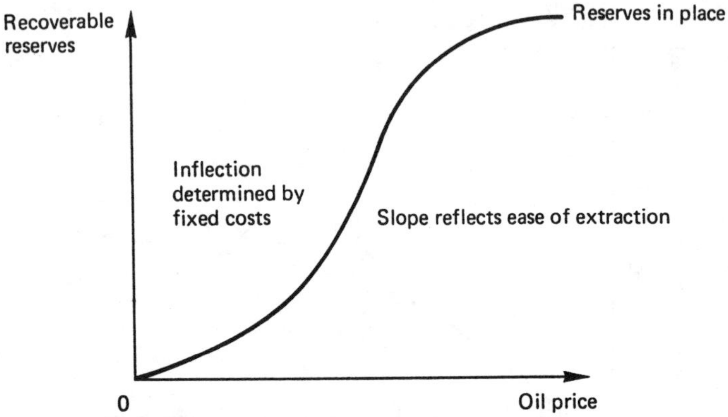

Given that our data is drawn from three basins we must introduce a pooled model:

$$R_i = L_i \, a_i \, e^{-b_i/Pe}$$

where the subscript i represents the East Shetland, northern North Sea and Moray Firth basins. Unfortunately, estimating six coefficients (three 'a's and three 'b's) cannot be justified with the above data. Although in tests each of the six t-statistics is satisfactory, incremental f-ratios show that incorporating a sixth coefficient does not significantly improve the regression. That is, data from the northern North Sea and from the Moray Firth basins do not vary enough to warrant separate estimates of two asymptotic and two growth parameters. We must omit either one asymptotic or one growth coefficient. In either case, f-ratios justify the inclusion of the other five variables. Studying the geology and formation size of the northern North Sea and Moray Firth basins suggests that asymptotic reserves are likely to be similar, while varying water depths and latitudes indicate that different growth parameters would be likely. This *a priori* reasoning is also supported by the f-tests which slightly favour the model with one asymptotic and two growth parameter dummy variables over the model with two asymptotic and one growth parameter dummy variable.

Commercial Considerations

Having established how to estimate asymptotic reserves, we now need to consider how rapidly this reserve level will be reached as oil prices rise in a commercial decision-making environment. The growth parameters estimated on the basis of economic supply prices reflect the price coefficients as if project decisions are taken with reference to costs but without reference to taxation.

Companies, however, are concerned about post-tax profitability. Specifically, we have assumed that once oil prices rise to yield revenues from the project sufficiently in excess of costs and taxes to cover the necessary risk premium (equal to the project's discounted capital costs) and to leave a real 5 per cent return, then the project will be declared commercial and may be included in our recoverable reserve estimate; hence the commercial (rather than the economic) supply oil prices are used to cumulate reserves (see the rightmost two columns of Table B.1). If the established real oil price exceeds a field's commercial supply price, our behavioural model suggests the representative firm will develop the field.

In this way, we recalculate our data and re-estimate the supply curve to take account of any tax system operating for North Sea companies. Ideally the tax system should only siphon off surplus profits from a field, as otherwise the allocation of resources will be distorted. If there were no distorting effect, the commercial supply price and the economic supply price for a given field would be identical. Most tax systems, however, are far from ideal and do affect the oil prices that are just necessary to induce development of projects (compare column two with column four of Table B.1 for the current tax injection). This distortion is explored in Chapter 4 but here we must recognise the significance of the tax system if our behavioural model is to represent the oil industry and produce accurate forecasts.

In performing this re-estimation to assess the impact of the current tax system it is worth pointing out that the commercial supply prices are calculated as if the current tax system had always been in operation. Since recoverable reserves will be influenced by future available returns, and not by the conditions applying under previous tax systems, data points generated by the current tax system only are relevant. Even though the profits available on a field such as Forties will, to a large extent, be determined by past tax systems, if we wish to forecast for the current tax system for new fields we need to know when Forties would just become commercial under the current tax system for new fields.

For a project such as Forties where production started in 1975, one must bear in mind that the tax system in 1975 was not the same as the tax system in 1984. For example, the bulk of Forties' capital costs will be 'upliftable' at a rate of 175 per cent. If we wish to use our model for forecasting, information based on an outdated tax system will not be relevant. Instead we should re-estimate the prices necessary to induce development, not only as if the price holds over the producing life of the field but also as if the current tax system holds over the life of the field. Clearly (for similar reasons to those in section 5) this price will be a poor guide to Forties' profitability; however, it would indicate the price necessary for development if another Forties were discovered today. As we wish to use our model to quantify the level of recoverable reserves from new areas, this latter price is relevant and we re-estimate the commercial supply prices on this basis.

As the tax system will not alter the asymptotic reserves in place, we must constrain our re-estimation to match the asymptotes estimated above. Hence we adopt a restricted (and weighted) least squares approach and construct the following structural equation:

$$R_i = L_i\, a_i\, e^{-(c_i/P_c)}$$

where a_i are the estimated asymptotes for the three basins, c_i are the commercial growth parameters (to be estimated), P_c are the commercial supply oil prices, that is the price where project revenues just outweigh all costs plus taxes.

The Quality of Each Licensed Block

The econometric accuracy of the estimates produced from this model supports the theory underlying the approach. The asymptotic form is a good approximation of the recoverable reserve/real oil price relationship. However, the model presented so far is only relevant to the area licensed in rounds one to four and it may not be possible to use it to forecast outside this area. In particular, we would estimate average recoverable reserves as if each area's average is invariant to the size of that area licensed. This implies that once a particular basin has been selected for licensing, either each block in that license is of uniform quality or the choice of which blocks to allocate first is made randomly. The former implication is of dubious merit since oil reserves tend to be concentrated in the centre of a block, and the latter implication does not correspond to government policy. The licensing history of each geological basin demonstrates that governments purposefully choose the most promising blocks first – the blocks generally towards the centre of a basin where potential oil-bearing structures are likely to be thickest – before licensing the less promising blocks, which are generally located nearer to the perimeter of a basin. Hence our database blocks will be better than typical and average recoverable reserves in the unlicensed areas of our database blocks are likely to be less than estimated in the above procedure.

Although the merits of this policy are not relevant here, it is important to ascertain whether this policy will continue in the future. For the moment this certainly seems to be the case; the oil companies have even been asked to nominate, from within broadly specified areas, the blocks they would like to see allocated in the seventh, eighth and ninth licence rounds, that is, the most promising of the remaining blocks.[13] A continued government policy of selecting the best blocks first is a reasonable *a priori* characteristic that is not yet contained in our model. If we fail to include this policy then our estimates will tend to overstate the reserves outside our database acreage.

Unfortunately we are restricted to using the experience from the first four licensing rounds which provides insufficient variance to be able to estimate such government intentions econometrically. Instead, government licensing policy must be modelled by an explicit specification of the rate of decline in the blocks licensed. In the above estimations we postulated that each block in a basin is identical. No less rigidly, we must now settle on a specification for the decline from the best block in the basin to the worst block in the basin.

For simplicity, we postulate that government policy constantly selects licences of decreasing merit. Although lack of information tends to cause the initial licensed blocks of an area not to be the best blocks when considered retrospectively, this possibility is dismissed. The first licences are then assessed to be the most promising, while the quality of later and later allocated blocks is assumed to deteriorate continually until the last block on the edge of the basin is licensed, which holds no recoverable reserves at any price.

Both our estimated parameters – the asymptote and the growth parameter – relate to the promise of a basin. However, only one parameter is likely to be affected by licensing policy. As the asymptotic reserves describe the underground rocks and as government policy tends to involve licensing the centre part of the basin first, then the level of asymptotic reserves for an incremental licence is likely to change with licence size. Conversely, the growth parameter is related to the geographical setting of the basin – for example, water depths and wind speeds – which will tend to be invariant over the whole site of a particular basin and will tend to be identical for each of that basin's blocks.

How quickly will the quality of each block decline? The relationship between each licensed block and the number of blocks already licensed may be represented by a function relating the extent of reserves in place (RP) to the size of the licensed area (L). Clearly, the first derivative will be positive, $\partial RP/\partial L \geqslant 0$; and the second derivative negative, $\partial^2 RP/\partial L^2 \leqslant 0$; reserves will only rise at a declining rate, so that marginal reserves from an additional licence will be greatest when few licences have been issued; but we may naturally assume something about the sign of a higher derivative also. The rate of decline will be likely to rise as the licensed area grows towards the total basin size, $\partial^3 RP/\partial L^3 \leqslant 0$, so the rate of decline will speed up as shown in Figure B.4. The trigonometric function of cosine describes the situation depicted in Figure B.4. In the absence of any conflicting evidence we adopt this formulation as an approximation to government policy. The intercept on the horizontal axis would be achieved when the entire basin has been licensed, and the intercept on the vertical axis represents the best block in that basin.

After integrating and suppressing the constant of integration (because the schedule must pass through the origin) the model specification becomes:

$$R_i = a_i \frac{2k_i}{\pi} \sin\left(\frac{L_i}{k_i} \frac{\pi}{2}\right) e^{-bi/p}$$

where the i subscript refers to each geological basin, a is the best asymptotic reserves, L is the number of blocks allocated, and K is the total basin size. In effect we have replaced L by $2k/\pi \sin\left(\frac{L}{K} \times \frac{\pi}{2}\right)$, and we have reinterpreted 'a', which instead of representing average reserves per block now signifies the best block of each basin.

As mentioned above, but more serious in this case, is the fact that the f-statistic restricts the number of dummy variables that can be usefully included. Separate dummies for the slopes and the intercepts of each basin are not justified, although one pair of dummy variables (one slope, one intercept) is statistically significant.

Figure B.4

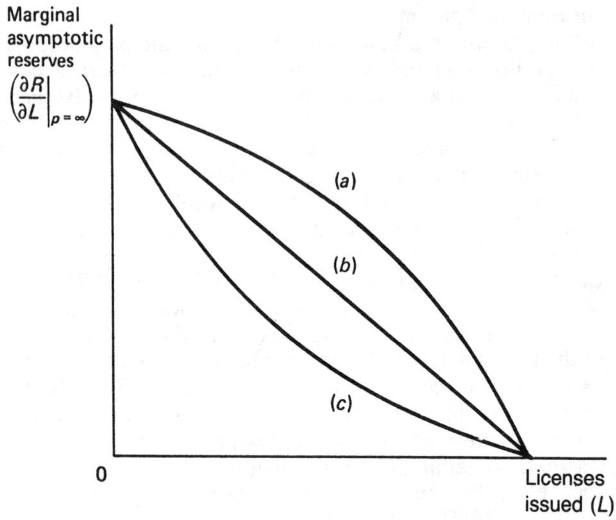

$$\partial^2 RP/\partial L^2 \big|_{p=\infty} \text{ is negative and}$$

$$\text{the rate of decline (a) increases } \frac{\partial^3 RP}{\partial L^3}\bigg|_{p=\infty} < 0$$

$$\text{(b) is constant } \frac{\partial^3 RP}{\partial L^3}\bigg|_{p=\infty} = 0$$

$$\text{(c) decreases } \frac{\partial^3 RP}{\partial L^3}\bigg|_{p=\infty} > 0$$

Geological Details Basin by Basin

A map of the basins can be found in the 1974 edition of the Brown Book (*Development of UK Oil and Gas Resources*). Table B.2, below, was derived by mapping the basins on to other maps and reading off the relevant details. In any exercise of this nature, the data is far from purely objective. Given the nature of the data's use, however, too great a degree of precision would imply spurious accuracy and is not necessary.

Model Estimates

Estimates from this model give the commercially recoverable reserves as a function of oil prices in a form suitable for forecasting. Summarising our results, the East Shetland basin has an estimated equation of:

$$R = \frac{85}{\pi} 2 \text{ Sin} \left(\frac{L}{85} \frac{\pi}{2} \right) 6.3668 e^{-27.5512/p}$$

Table B.2 Brief geological and geographic details of each basin

Basin	*Geological legend*	*Water depths* (metres)	*Basin size* (blocks)
East Shetland	Jurassic/Triassic/ Cretacious	100–200	165
Northern North Sea	Triassic/Jurassic/ Cretacious/Tertiary	50–100	85
Moray Firth	Permian/Mesozoic	25–200	134
Forth Approaches	Lower Triassic	50–100	74
Fair Isle and Orkney Isle	Pre-permian	100–200	19
Faeroes (Trough)	Permian/Mesozoic/Tertiary	200–1000	51
West Shetland	Permian/Mesozoic/Tertiary	100–200	77
West Orkney	Mesozoic/Tertiary	50–100	30
North Minch	Permian/Mesozoic/Tertiary	100–200	29
Sea of Hebrides (Trough)	Tertiary/Mesozoic/Permian	100–200	32
Inner Hebrides (Trough)	Tertiary/Mesozoic/Permian/ Pre-Cambrian	50–100	18
Rockall (Trough)	Permian/Mesozoic/Tertiary	1000+	486
Rockall – Hatton	Permian/Mesozoic/Tertiary	200–1000	114
Irish Sea	Permian/Mesozoic	25–50	47
Southern Sub	Tertiary/Cretaceous	50–100	65
South-western Approaches	Cretaceous/Tertiary	100–200	64
West English Channel	Cretaceous/Permo-Triassic	50–100	58
East English Channel	Cretaceous/Tertiary/Jurassic	50–100	56
Unst	Pre-Permian	100–200	6
Northern Sub	Tertiary	50–100	13
Cardigan Bay	Permian/Mesozoic	25–50	7
Clyde	Pre-Cambrian to Tertiary	25–200	13
Bristol Channel	Permian	25–50	15
Southern North Sea	Paleozoic/Mesozoic	25–50	—

The northern North Sea basin has an estimated equation of:

$$R = \frac{165}{\pi} 2 \sin\left(\frac{L}{165} \; \frac{\pi}{2}\right) 3.8887e^{-7.8162/p}$$

The Moray Firth basin has an estimated equation of:

$$R = \frac{134}{\pi} 2 \sin\left(\frac{L}{134} \; \frac{\pi}{2}\right) 3.8887e^{-7.8162/p}$$

Results from these equations are compared to our database in Figure B.2.

5 FORECASTING PROCEDURE

We have constructed three supply curves, one for each of the three regions with relatively well-explored histories. These estimates form the basis for the forecasts in Chapter 4. Leading up to our final type of forecast, the forecasting procedure may be broken down into three segments: extrapolating the reserves which, if oil prices are higher than those of our database, will be commercially recoverable from the areas licensed in the first four rounds of each of our three base areas; extrapolating the relationship between recoverable reserves and oil prices to the area of each of the three base regions that was not licensed in the first four rounds; and finally, extrapolating this relationship to the other geological basins of the UK. Further uncertainty and more speculative ideas are added as we move from the first to the third type of forecast.

Long-Run Oil Prices

The essence of our model is to predict the reserves which will eventually be recovered from the UK which will only be known with certainty when the last offshore oil field has been depleted. One part of these reserves may come from the acreage of the East Shetland, the northern North Sea and the Moray Firth basins that was allocated in the first four licence rounds. The estimated equations given in the previous section allow this forecast to be made conditional on the eventual long-run price of North Sea oil. This price will reflect the increasing scarcity of oil worldwide and will be a function of long-run supply and demand considerations.[14] Here we are not interested in short-term tendencies which might raise oil prices above, or lower oil prices below, their long-run values.[15] An oil – field development decision is a long-run plan that will be assessed on long-run criteria. The long-run oil price will be constrained in world markets by the price of an alternative energy source that can be a commercial replacement to oil in its most valuable use.

If the eventual oil price were to be lower than current expectations for the established oil price, then our supply curve would not accurately reflect the fall in commercial reserves. The supply curve would ignore the capital costs which have already been sunk on many fields. Given these sunk costs, the prices necessary for *continued* development fall dramatically.[16] Even if oil prices were to drop by large amounts, most fields with sunk capital costs would still look commercial on *avoidable* costs and production would continue. However, our supply curve is estimated on the prices necessary to ensure commercial production over the life of each field, hence the estimated fall in supplies would suggest recoverable reserves will decline more rapidly than the actual fall in reserves. In principle, we should overcome this problem by re-estimating the downward segment of our supply curve on the basis of avoidable costs only and then construct a kinked supply curve around current established oil prices. In practice this problem will be avoided so long as the eventual long-run oil price is never exceeded by expectations of the established oil price. We must be careful, however, not to forecast with prices much below $20 per barrel without adapting our model.

Unfortunately the eventual long-run oil price cannot be predicted with any degree of certainty. Not only must we predict long-run substitution costs, but as our dependent variable is expressed in deflated values we must also assume the paths for UK inflation relative to a base currency are correctly anticipated by our representative firm. This prediction has been made by projecting various paths for nominal dollar oil prices into the next century, and then deflating the value back into 1983 terms and adjusting for exchange rate movements. Whether long-run oil prices will have peaked by the year 2000 is not certain; however, a 30-year period is probably long enough for the transition to a non-oil based economy.

At one extreme, recent oil prices of around $20 per barrel (in real 1983 terms) may be an estimate of the long-run oil price. Given that realistic substitutes for oil are estimated to cost as little as $20 to $30 per barrel equivalent,[17] eventually commercial oil reserves recovered from the areas of the East Shetland, the northern North Sea and the Moray Firth basins licensed in the first four rounds may not be much greater than the total reserves used in the database. The chances are that real long-run oil prices will be greater than this estimate, perhaps as high as $50 per barrel. This price would induce considerably more recoverable reserves from the three database areas than used in the database. Supporting this estimate, the Department of Energy have suggested $44 per barrel (in 1980 terms) as likely by the year 2000.[18]

Unlicensed Areas in Database Basins

The modified specification of section 4 permits our model to forecast recoverable reserves in areas of the East Shetland, northern North Sea and Moray Firth basins that were not licensed in the first four licence rounds. The East Shetland and northern North Sea basins seem unlikely to provide further substantial reserves from additional licences, but the Moray Firth basin seems more promising.

Other Basins

In addition, there are other potentially oil-bearing basins on the UKCS apart from the three basins we have considered so far. Recoverable reserves from these basins are likely and our final forecast of UKCS reserves must include an estimate of their size. To facilitate this estimate we wish to construct similar supply relationships to those already developed for the East Shetland, northern North Sea and Moray Firth basins. For each potential oil-bearing basin, both the asymptotic reserves in place for the most promising block and the growth parameters reflecting the speed with which the asymptote is reached must be assessed. As yet only negligible areas of these basins have been licensed and the absence of discoveries precludes any direct quantitative (econometric) assessment. Instead we adopt a more qualitative approach, assessing crucial features of each basin to see if any particular basin is similar or dissimilar to the three basins we have assessed and quantified directly. This information is drawn together in Table B.2. Such a qualitative

Table B.3 Oil price reserve relationship in other basins

Basin	*Commercial growth parameter*	*Asymptotic reserves in best block*
East Shetland	c_3	a_3
Northern North Sea	c_2	a_2
Moray Firth	c_4	a_4
Forth Approaches		Geology not promising
Fair Isle and Orkney Isle		Poor geology/basin size insufficient
Faeroes (Trough)	$1.1c_3$	$\frac{150}{170}\frac{5}{6}a_3$
West Shetland	c_3	$\frac{77}{170}\frac{5}{6}a_3$
West Orkeny		Small basin with unpromising geology
North Minch	c_3	0
Sea of Hebrides	c_3	$\frac{32}{170}\frac{5}{6}a_3$
Inner Hebrides		Basin too thin
Rockall (Trough)	$5c_3$	$\frac{5}{6}a_3$
Rockall–Hatton	$1.1c_3$	$\frac{150}{170}\frac{5}{6}a_3$
Irish Sea		Small basin with unpromising geology
Southern Sub	c_2	$\frac{65}{170}\frac{2}{6}a_3$
South-western Approaches	c_3	$\frac{150}{170}\frac{2}{6}a_3$
West English Channel	c_2	$\frac{90}{134}\frac{2}{5}a_4$
Eastern English Channel	c_2	$\frac{90}{170}\frac{3}{6}a_3$
Unst		Small basin: unpromising geology, deep, harsh waters
Northern Sub		Geology not promising
Cardigan Bay	c_4	0
Clyde		Small basin with unpromising geology
Bristol Channel		Small basin with unpromising geology
Southern North Sea		Gas-bearing rocks

Note: The two fractions for the asymptote are first, the proportionate size of the basin, and second, the geological similarity of the new basin to the database basin: see Table B.1.

approach cannot produce meaningful unique estimates for each basin, but we hope it can produce values for the oil price reserve relationship in total which are likely to be close to the true values of the relationship. More importantly our principal concern here is not to forecast recoverable reserves in total but to ascertain the sensitivity of reserves to economic parameters such as oil prices or tax systems. To this end the inaccuracies in our estimates in absolute terms are not relevant.

Some 24 basins have been identified on the UKCS South of the 62°N parallel and are listed in Table B.3. For each basin the water depths and local climates have been compared with those experienced in the East Shetlands, northern North Sea and Moray Firth basins, and (rather tentatively) coefficients have been assigned to each region's commercial growth parameter. These are expressed in the table as a fraction of the database basin's parameters. While this is a fairly heroic process, at least the determinants of the growth parameters (wind speeds, water depths, wave heights, and so on) are well documented.[19] The determinants of the asymptote are largely unknown, so that assigning coefficients to the asymptotic parameter is even more speculative. It follows that parameters should only be quantified to a rough order-of-magnitude precision. The table shows our central case estimate where this has been assigned, or some comment to indicate why a basin has been ignored. The background to these estimates is based on geological, geographic and industry information but will not be dwelt on here: the methodology is highly subjective and would differ between analysts. These are our best 'guestimates' and we merely hope that whatever errors remain might cancel. They are expressed in the table as a fraction of the database basin's asymptotes.

Calculating each basin's commercially recoverable reserves at a given long-run oil price and aggregating over all the basins yields an estimate of the commercially recoverable oil reserves from the entire UKCS south of the 62°N parallel.

6 ECONOMETRIC RESULTS

Forecasts from our model are shown in Table B.4 for a range of oil prices. The calculations demonstrate how sensitive recoverable reserves are to long-run oil prices: imputed elasticities for total UK reserves area around 0.8.

Table B.4 UK recoverable oil reserves (million tonnes)

	Real long-run oil price (1983 values, $ per barrel)				
	30	35	40	45	50
Central estimate	3362	3805	4213	4595	4955

7 MODEL WEAKNESSES

Before we use the model to quantify the impact of North Sea taxation there are several weaknesses in our approach that should be exposed. We have already stressed the weakness of our model to forecast lower supplies than currently under development. Other weaknesses fall into one of two categories, either methodological or restrictive. The methodological weaknesses are more serious and are considered first.

Methodological Limitations

Methodological weaknesses involve problems which cannot be resolved by sensitivity tests but which the model has inherent difficulties in tackling. Two main problems should be mentioned: the problem of technological change and the problem of producers amending output profiles in response to price changes. The effect of technological advances is to shift supply curves to the left. Given that our model involves no technical parameters and is timeless, our forecasts will increasingly underestimate recoverable reserves. In an activity where the engineering and construction processes are at the forefront of technology, this omission may be serious. This is a problem implicit in any cross-sectional study; the data is based on the technology of today and may not be relevant for forecasting into times when the technology will have advanced. Figure B.5 shows the estimate of C, based on the data points between A and B, to underestimate reserves if technology has shifted the supply curve from SS to S^TS^T.

In our application of cross-sectional data, however, this problem may not be too acute. Several fields in our simulated database are using the technology of a few years ago, since this is when development on these fields began. Our data points are not on the curve SS of Figure B.5 which is repeated in Figure B.6, but are from a group of Curves S^NS^N, reflecting earlier technologies. Conveniently for us, technology has advanced as oil prices have risen during the period 1970–80. It follows that the data points will not fall randomly on the curves S^1S^1 . . . S^NS^N, but will be in an ordered sequence moving to a higher curve as oil prices rise, such as shown by the crosses on Figure B.6. This yields the estimated supply curve EE. Hence so long as technology improves as established oil prices rise – and there is probably a causal link from the latter to the former – then forecasting on the basis of EE will present us with an accurate estimate of C^1. Although it would be desirable to separate our forecasts into a component involving a move along a curve (caused by price changes) and a component involving a shift of a curve (technological advance), this separation should not markedly affect the accuracy of our forecasts.

The second methodological weakness concerns the way the model ignores the ability of producers to vary field output profiles in response to price changes. Our cross-sectional data is based on information about output profiles chosen with today's oil prices. We interpolate at lower oil prices whether these fields would still be commercial by considering if our representative firm would develop them and for each field the decision must be either

Figure B.5

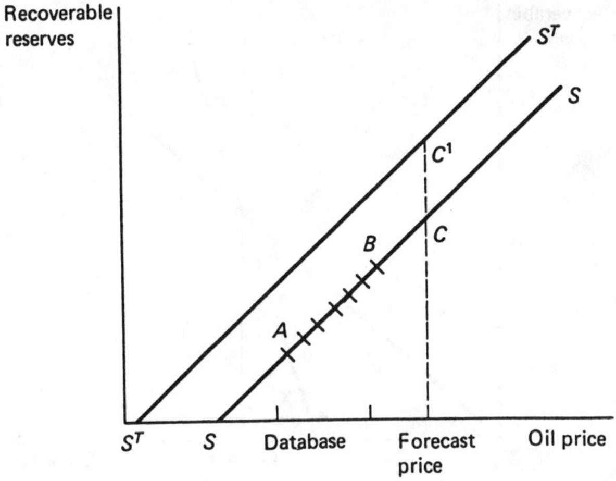

Figure B.6

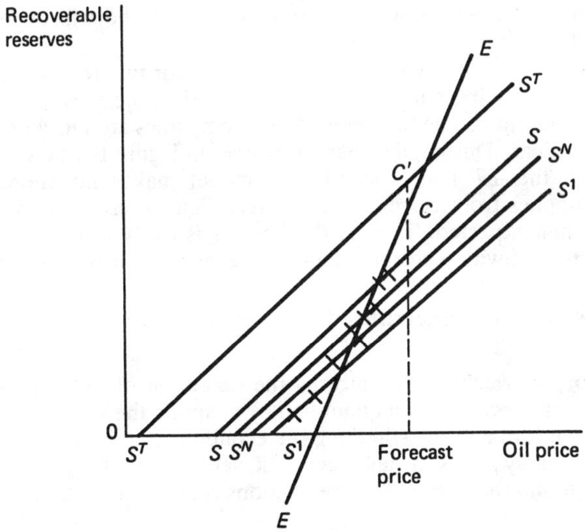

xx = Data points

to develop or not to develop. In practice, however, companies will have another option; that is, to reduce the scale of the project and only produce the most profitable barrels. Hence our simulated database will be a series of rather pessimistic points, such as the crosses in Figures B.7, below the true

Figure B.7

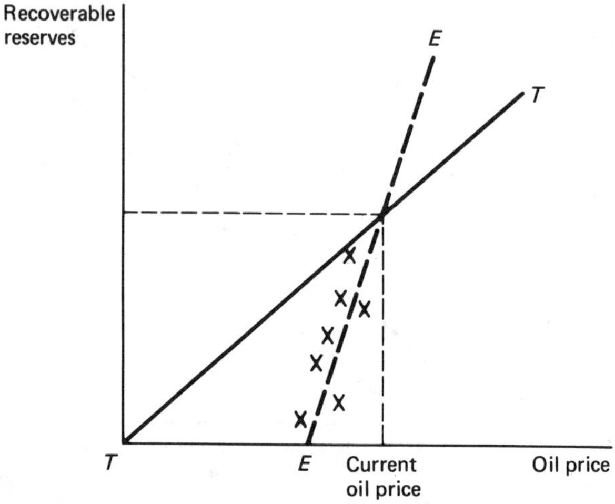

recoverable reserves along the line *TT*. Our estimated line, *EE*, will overestimate recoverable reserves above current oil prices, and will overstate the long-run supply elasticity.

The extent of this bias may not be too serious for two reasons. First, the development decision is a long-run decision based on the *established* price of oil. Once an output profile has been chosen companies are fairly reluctant to change their ideas. This implies that the crosses in Figure B.7 may not be too distant from the *TT* line. Second, our model makes no allowance for increased supplies from currently commercial fields if established oil prices are higher than expected. That is, the line *TT* is likely to be kinked about current oil prices, with a steeper slope to the right of current oil prices.

Model Restrictions Concerning the Representative Firm

This category of weaknesses concerns the definition of our representative firm. Each weakness has been quantified by adapting the specification of the representative firm and re-estimating the economic supply prices and/or the commercial supply prices to yield new coefficients in an otherwise unchanged model. We found that none of the restrictions considered materially affected our results.

First, we have respecified the return that is just necessary to induce our representative firm to bear the risks and proceed with development. The basic model assumed the level would just yield a normal return on capital once the real return exceeded 5 per cent, plus an allowance for risk based on the discounted value of each field's capital costs. This respecification proceeded along two dimensions. In re-estimating the model both the 5 per cent rate of return and the risk premium specification were varied.

We also adjusted the return necessary to induce development in the absence of taxes relative to the return necessary with taxation included. This adjustment may be necessary for two reasons: one concerning the normal return and one the risk premia. If we believe that in the absence of taxes throughout the economy a higher return to capital would be available, then a higher return would be necessary in the absence of taxation than with taxation included. The economic supply prices are used to find the asymptotic level of reserves, a level which is defined by nature rather than by artificial tax arrangements implemented by mankind, and even if via taxation the normal rate of return declines the asymptotic reserves should not be affected. Thus a higher rate of return may be used in the economic supply price calculations than in the commercial supply price calculations. For reasons connected with the risk premia, however, we may wish to use a lower rate of return in the economic than in the commercial supply price calculations. If the additional profits taxation levied on North Sea companies generates extra uncertainty, then a higher risk premium would be required with taxation included.

The second type of respecification concerns the nature of the tax system. Some companies' corporate profits present them with an ability to use offshore costs to defer other profits, thus enhancing the commercial attractiveness of offshore ventures. We have allowed for this possibility and re-estimated our model as if our representative firm can apply this tax advantage to all fields. Although use of this tax advantage may become increasingly common, it has not been sufficiently widely available for our model to use it as a base for forecasting and this model is not the focus of our attention. Meanwhile, the significance of this tax advantage is clear.

A third respecification involves our representative firm's expectations on the growth of oil prices. We have assumed the established oil price is considered as a constant real value. Perhaps it should be considered as a constant nominal value. Conversely, perhaps it may be viewed as a real value with a steady rate of growth. Variations on both views have been analysed. The latter view, however, would be conceptually inconsistent. If real oil prices are rising, then a field with constant real costs would become increasingly profitable the longer it is delayed: the reward from delaying will be additional revenues with no additional costs, and the incentive to delay development will last forever. Our model would suggest that no fields will ever be pursued. The notion that a field's profitability is so closely related to the timing of its development runs against the cross-sectional nature of our model, and hence should not be considered further.

8 USE OF MODEL

Our model is now ready to quantify the effects of the different components of the tax system and the impact of the tax dynamics. For each specification of the tax system, the commercial supply price is recalculated (as explained in section 3), the commercial supply equation is re-estimated econometrically (see section 4), and the unknown geology is reassessed on the basis of the re-estimated coefficients (see section 5). The quantifications are presented in Chapter 4 and then the model is used in Chapter 5 to examine the supply curve under PPT.

Notes and References

1 Area for Concern

1. Select Committee on Energy, *North Sea Oil Depletion Policy*, Session 1981–2, Third Report, House of Commons Paper 337 (HMSO, London, May 1982).
2. Select Committee on Energy, *Press Statement: North Sea Oil Depletion Policy* (HMSO, London, 18 May 1982).
3. Select Committee on Energy, *North Sea Oil Depletion Policy*, para. 68; see also paras 72 and 73.
4. Secretary of State for Energy and the Chancellor of the Exchequer, *Memorandum on the Select Committee on Energy Report North Sea Oil Depletion* (HMSO, London, 31 August 1982).
5. Ibid., para. 23.
6. Ibid., para. 32.
7. A. Kemp and D. Rose, 'How to Keep Oil Flowing when the Price Moves Down', *Financial Times Energy Economist*, 15, April 1983.

2 Government Policy: Targets and Instruments

1. The geological forces that alter costs are explained in P. Davidson 'The Economics of Natural Resources', *Challenge*, 22, 1 (March/April 1981) and in F. W. Mansvelt Beck and K. M. Wigg, *The Economics of Offshore Oil and Gas Supplies* (Lexington, Toronto, 1977).
2. M. A. Adelman *The World Petroleum Market* (Johns Hopkins, London, 1972) elaborates on these points (particularly pp. 19–21).
3. The extension to a larger number of periods is straightforward.
4. Some have argued that reservoir damage occurs if too fast an initial rate of production is pursued, and thus overall recovery may be reduced. For example, see R. S. Pindyck 'The Optimal Exploration and Production of Non-recoverable Resources', *Journal of Political Economy*, 10, 2 (October 1978). Our conceptualisation does not allow for this possibility. By linking the capacity level to the rate of early production this interaction could be included in our model. We doubt, however, the validity of these arguments in a North Sea context, where the range of feasible initial production rates is technically and geologically limited.
5. First discussed by H. Hotelling, 'The Economics of Exhaustible Resources', *Journal of Political Economy*, 39 (1931).
6. Adelman, *The World Petroleum Market.*
7. More details can be found in C. Robinson and J. Morgan, *North Sea Oil in the Future: Economic Analysis and Government Policy* (Macmillan, London, 1978).
8. See T. Barker and V. Brailovsky, *Oil or Industry? Energy Industrialization and Economic Policy in Canada, Mexico, Norway the Netherlands and the United Kingdom* (Academic Press, London, 1981).

9. Other schemes of payment (for example, royalty payments or net profit arrangements) link the bids to performance in later years, but these are not considered here. They not only alter the distribution of windfall profits but – for the same reasons that CT tends not to operate neutrally – they also alter the profit-maximising depletion and capacity choices. The other schemes may thus be thought of as lower cash payments and greater CT liability, with a bidded CT rate. See S. L. McDonald, *The Leasing of Federal Lands for Fossil Fuels Production* (Johns Hopkins, London, 1979) and J. B. Ramsey, *Bidding and Oil Leases* (Jai Press, Connecticutt, 1980).
10. See E. J. Mishan, *An Introduction to Normative Economics* (George Allen & Unwin, London, 1980).
11. And as UK oil consumption has dropped towards self-sufficiency. See the statements by then Secretary of State for Energy, Mr David Howell, reported in Department of Energy, *Development of the Oil and Gas Resources in the United Kingdom* (HMSO, London, 1981).
12. This corollary is due to K. W. Dam, *Oil Resources: Who Gets What How?* (University of Chicago Press, London, 1978).
13. In particular, with UK nationalised corporation involvement.
14. The Varley Guidelines appeared in HMSO, *Weekly Hansard (Commons),* vol. 882, columns 648–50, (Written Answers) and the Howell statement in Department of Energy, *Development of the Oil and Gas Resources*. Most recently, Mr Lawson reassured the industry that no oil production cutbacks would be imposed 'at least until the end of 1984'; HMSO, *Weekly Hansard (Commons)*, Volume 25, Issue No. 1246, column 6, (Written Answers).
15. C. Robinson, 'The Errors of North Sea Policy', *Lloyds Bank Review*, July 1981.
16. Note that the legal restraints on collusion over work programmes will tend to be less restrictive than legal restraints on collusion over price bids.
17. Although not included in Mr Errol's five stated preferences used in discretionary licensing, this has been used as a criteria for allocating licences; see Dam, *Oil Resources*.
18. See P. J. Lehmann, 'Free Market in British Gas?', *Economic Affairs*, January 1982, and P. J. Lehmann and N. R. Shaw, 'British Gas Buying Rights and Natural Gas Depletion', *Energy Policy*, 10, 1 (March 1982).
19. This is well documented by C. Robinson, *Competition for Fuel* (Institute for Economic Affairs, London, 1971) and C. Robinson and C. Rowland, 'North Sea Oil and Gas', in P. Johnson (ed.), *The Structure of British Industry* (Granada, London, 1980).
20. See the references in note 14. This may no longer be true since the BGC suggested that all companies which find gas will receive at least 22p per therm.

3 The Economics of Oil Taxation

1. R. F. Hayllar and R. T. Pleasance, 'UK Taxation of Offshore Oil and Gas' (B. Butterworths, London, 1977), p. 72.

2. Wilson Committee Report on the functioning of the financial Institutions, *The Financing of North Sea Oil*, Research Report No. 2 (HMSO, London 1980).
3. As introduced by R. A. Musgrave and P. B. Musgrave, *Public Licence in Theory and Practice* (McGraw-Hill, London, 1976).
4. HMSO, *Weekly Hansard (Commons)*, Volume 880/887, Issue No. 990, columns 295–6.
5. Ibid., column 291.
6. HMSO, *Weekly Hansard (Commons:)*, Volume 888, Issue No. 993, column 1654.
7. HMSO, *Weekly Hansard (Commons)*, Volume 886/887, Issue No. 990, column 291.
8. J. Morgan, 'The Promise and Problems of PRT', IFS Conference Paper, January 1976, highlighted the stormy passage of PRT from first Bill to Act.
9. HMSO, *Weekly Hansard (Commons)*, Volume 955, Issue No. 1118, columns 753–4.
10. Ibid., column 755.
11. HMSO, *Weekly Hansard (Commons)*, Volume 968, Issue No. 1140, columns 254 and 255.
12. HMSO, *Weekly Hansard (Commons)*, Volume 980, Issue No. 1166, columns 1464 and 1465.
13. Ibid., column 1497.
14. HMSO, *Weekly Hansard (Commons)*, vol. 994, Issue No. 1187, column 317.
15. Ibid., column 320.
16. HMSO, *Weekly Hansard (Commons)*, Volume 1000, Issue No. 1199, column 771.
17. HMSO, *Weekly Hansard (Commons)*, Volume 886/887, Issue No. 990, column 1660.
18. HMSO, *Weekly Hansard (Commons)*, Volume 19, Issue No. 1235, column 745.
19. Ibid., column 745.
20. Ibid., column 745.
21. HMSO, *Weekly Hansard (Commons)*, Volume 25, Issue No. 1247, column 109.
22. HM Treasury, Press Release, 9 June 1981.
23. HMSO, *Weekly Hansard (Commons)*, Volume 38/39, Issue No. 1272, column 148.
24. HMSO, *Weekly Hansard (Commons)*, Volume 56, Issue No. 1303, column 300.
25. Much work at the University of Aberdeen, Department of Political Economy – notably by Alexander Kemp – has centred on detailed analyses of various economic issues associated with the taxation of North Sea oil. Recent publications have studied the taxation implications of incremental investment in the North Sea and the handling of oil field abandonment. Considerable and on-going uncertainty exists as to the timing and the nature of any tax concessions to be introduced in order to encourage incremental investment and as to any system of fiscal relief for

abandonment costs. These would not be issues with a simple system of rent taxation so outlined in Chapter 5. See A. Y. Kemp and D. Rose, *Fiscal Aspects of Incremental Investments in the UK Continental Shelf*, North Sea Study Occasional Paper No. 20 (Department of Political Economy, University of Aberdeen, December 1984), and A. Y. Kemp and D. Rose (assisted by B. Creed), *Fiscal Aspects of Field Abandonment in the UKCS*, North Sea Study Occasional Paper No. 22 (Department of Political Economy, University of Aberdeen, October 1985).

4 The Impact of Taxation on Recoverable Reserves

1. We have inverted the usual supply curve and plot supply on the vertical axis against price on the horizontal axis: hence the producer surplus is the area under a supply curve rather than the are above a supply curve.

5 Petroleum Profits Tax: An Alternative Proposal

1. IFS Committee, 'The Taxation of North Sea Oil' (IFS, London, December 1981).
2. Publication of the IFS proposals was reported in the *Financial Times*, *The Times*, the *Daily Telegraph* and the *Guardian* on 22 December 1981.
3. Strictly, asking why the proposals were not implemented at the start of 1982 is not identical to asking if they should be implemented at some later date. However, we believe the false perceptions which predominated at the start of 1982 have yet to change.
4. Details may be found in IFS Committee, 'The Taxation of North Sea Oil'.
5. A. J. Clunies Ross, 'North Sea Oil and Gas Taxation: A Case for Reform', *Three Banks Review*, 134 (June 1982), p. 36.
6. K. W. Dam, *Oil Resources: Who Get What How*? (University of Chicago Press, London, 1976)
7. E. Cary Brown, 'Business – Income Taxation and Investment Incentives', in L. A. Merzles et al., *Income Employment and Public Policy – Grays in Honor of Alvin H. Harsen* (Norton, New York, 1948) pp. 300–16
8. W. Mayo, 'Rent Royalties', *Economic Record*, 55 (September 1979), pp. 202–13.
9. R. Garnaut and A. J. Clunies Ross, 'Uncertainty, Risk Aversion and the Taxing of Natural Resource Project', *Economic Journal*, 85 (1975), pp. 272–87.
10. Mayo, 'Rent Royalties'.
11. K. F. Palmer, 'Mineral Taxation Polices in Developing Countries: An Application of Resource Rent Tax', *IMF Staff Papers*, 27 (1980), pp. 517–42.
12. R. Garnaut and A. J. Clunies Ross, 'The Neutrality of the Resource Rent Tax', *Economic Record*, 55 (September 1979), p. 193.
13. Ibid., pp. 193–201.

14. M. T. Sumner, 'Progressive Taxation of Natural Resource Rents', *Manchester School*, 46 (March 1978), pp. 1–16.
15. M. P. Devereux and C. N. Morris, '*North Sea Oil Taxation*', IFS, Report Series No. 6 (London, December 1983).
16. Sumner, 'Progressive Taxation'.
17. Clunies Ross, 'North Sea Oil and Gas Taxation'.
18. Palmer, 'Mineral Taxation Policies'.
19. Clunies Ross, 'North Sea Oil and Gas Taxation', p. 43.
20. In the past, since the existing tax system does not and never has exhibited this feature, government has had necessarily to adjust the tax rates and allowance. See also C. Johnson, 'North Sea Energy Wealth 1965–1985', *Financial Times*, 1979, on the lack of automatic flexibility.
21. See HM Treasury, Press Release, 9 June 1981.
22. See Chapter 3 where we suggest that a second-best criterion after progressive average tax takes is progressivity in marginal tax takes.
23. The rate of return is given by in the formula:

$$\frac{(1 + t1)^{L1}}{(1 + q)^{L1}} \times r1 + \frac{(1 + t2)^{L2}}{(1 + q)^{L2}} \times r2 + \frac{(1 + t3)^{L3}}{(1 + q)^{L3}} \times r3 = 0$$

Where $t1$ = threshold rate of return in band 1, $L1$ = lag before threshold on band 1 is passed, and $r1$ = rate of tax in band 1.

24. The tax reliefs on capital and operating costs are usually quoted as 101 per cent and 88 per cent respectively (allowing for some cost relief in royalty – as permitted under the first four licence rounds – increases these figures to 102 per cent and 89 per cent respectively). A. Kemp and D. Cohen, *The Impact of the System of Petroleum Taxation in the UK on Oil Operations and Government Revenues*, Fraser of Allander Institute Research Monograph No. 8 (University of Strathclyde, 1980), discuss these reliefs and show how the reliefs have been altered by PRT changes. However, with the long lead times encountered offshore the companies will be concerned with the timing of the reliefs relative to the timing of the cost outlays, a concern not incorporated in undiscounted values of the reliefs. Looking at the discounted values of the reliefs undermines the tax incentive to invest in capital relative to the tax incentive on operating expenses. Since tax assessable profits are likely to be positive when operating costs are committed, the operating cost relief will be claimed only one year late (as taxes are paid late). The capital relief, however, may not come until anything up to ten years after the capital expense, though on average it will be claimed two to five years after the expenditure. Depending on the rate of discount the operating cost tax relief will be around the low 70s per cent. The capital cost tax relief is likely to be much lower, possibly as low as 30 per cent but may be up to the mid-70s per cent depending not only on the rate of discount but also on the length of the lag. If companies are concerned with these discounted reliefs then the distortionary intra-industry impact on the Shelf may not be alleviated.
25. IFS Committee, 'The Taxation of North Sea Oil', para. 33.

6 A Desirable Solution

1. See Appendix D, IFS Committee, 'The Taxation of North Sea Oil' (IFS, London, December 1981), which details the necessary legislative changes. The changes are, surprisingly, few.
2. Ibid., para. 33.
3. See HM Treasury, Press Release, 9 June 1981.
4. For example, consider the lengthy legal process initiated whenever Southern Basin gas-producing companies seek to establish events have changed sufficiently to necessitate a change in the price contracted with the BGC.
5. UKOOA, *Submission to the Chancellor of the Exchequer on the 1981 Changes to the United Kingdom Oil Tax Regime* (UKOOA, London, October 1981); Brindex, *Oil Taxation in the United Kingdom: Submission to the Chancellor of the Exchequer* (Brindex, London, September 1981); and 'Taxation Threatens 30 Oilfields'; *Financial Times*, 22 December 1982, which reports unchanged industry views.
6. UKOOA, *Submission to the Chancellor of the Exchequer*, p. 5.
7. Brindex, *Oil Taxation in the United Kingdom*, p. 1.
8. Ibid., p. 17.
9. Ibid., p. 10.
10. UKOOA, *Submission to the Chancellor of the Exchequer* p. 15.
11. Wilson Committee Report on the Functioning of the Financial Institutions, *The Financing of North Sea Oil*, Research Report No. 2 (HMSO, London, 1980).
12. Since then, the 1983 two-tier tax system was designed to encourage new marginal developments without affecting established projects which have the greatest impact on the total tax level.
13. Not all of the benefits to the private sector would be net benefits: some benefits of turning otherwise uncommercial discoveries into valuable assets must be used to compensate the companies whose existing fields suffer in the switch to PPT. In this way the costs of negotiating and agreeing compensation would be bypassed, and no explicit compensation would need to be arranged.
14. See C. Rowland, 'North Sea Tax Changes: The Missed Opportunity', *Noroil*, May 1982.
15. For instance, exploration expenditure need not be deemed abortive to be deductible from PRT; and PRT relief could be claimed for shared assets such as pipelines.
16. Inland Revenue, Press Release, *Oil Taxation* (15 March 1983).

Appendix B How Much Oil is There?

1. The earliest articles are those of L. C. Gray, 'Rent Under the Assumption of Exhaustibility', *Quarterly Journal of Economics*, 28 (1914), and H. Hotelling, 'The Economics of Exhaustible Resources', *Journal of Political Economy*, 39 (1931). See also the following reviews: O.C.

Herfindahl, 'Depletion and Economic Theory', in M. Gaffrey (ed.), *Extractive Resources and Taxation* (University of Wisconsin Press, Madison, Wisconsin, 1967); R. M. Solow, 'The Economics of Resources or the Resources of Economics', *American Economic Review*, Papers and Proceedings (1974); and P. S. Dasgupta and G. M. Heal, *Economic Theory and Exhaustible Resources* (Cambridge University Press, 1979). R. S. Pindyck, 'The Optimal Exploration and Production of Non-renewable Resources', *Journal of Political Economy*, 10, 2 (October 1978), and V. L. Smith, 'Economics of Production from Natural Resources,' *American Economic Review*, 58 (June 1968) are particularly helpful.

Articles of a more applied nature appear in C. Robinson and J. Morgan, *North Sea Oil in the Future: Economic Analysis and Government Policy* (Macmillan, London, 1978); R. Lecomber, *The Economics of Natural Resources* (Macmillan, London, 1979); D. Evans, *Western Energy Policy: The Case of Competition* (Macmillan, London, 1978); and T. Barker and V. Brailovsky, *Competition* (Macmillan, London, 1981). A recent resumé of attitudes towards depletion appears in the Select Committee on Energy, *North Sea Oil Depletion Policy*: Minutes of Evidence and Report (HMSO, London, 1981).

2. D. C. Ion, 'Classification of Proved Reserves' and 'Classification of Additional Resources', *Petroleum Economist* (December 1979 and January 1980).
3. An excellent introduction to the geology of oil can be found in 'Our Industry Petroleum' (British Petroleum, London, 1970).
4. Department of Energy, *Development of the Oil and Gas Resources of the UK 1984* (HMSO, London, 1984).
5. P. L. Eckbo, H. D. Jacoby and J. L. Smith, *Oil Supply Forecasting: A Disaggregated Process Approach* (MIT, February 1977); P. L. Eckbo, *The Supply of North Sea Oil* (MIT, July 1977); C. J. Johnson, 'North Sea Energy Wealth 1975–2000', *Financial Times*, L. 1979; P. Odell and K. Rosing, *Optimal Development of North Sea's Oilfields* (Kogan Page, London, 1977); and UKOOA, *Exploration and Development of UK Continental Shelf*, Oil Energy Commission Paper No. 17 (HMSO, London, 1978), amongst others.
6. D. R. Basu, *Future Energy Policies for the UK* (Macmillan, London, 1981) reviews the literature in Chapter 2; see also F. M. O'Carrol and J. L. Smith, *Probabilistic Methods of Estimating Undiscovered Petroleum Resources* (MIT, January 1981).
7. The specification and quantification of this return is detailed below.
8. J. Mitchel, 'Taxation of Oil and Gas Revenues of Four Countries: the United Kingdom', *Energy Journal*, April 1982; A. Kemp and D. Cohen, *The Impact of the System of Petroleum Taxation in the UK on Oil Operations and Government Revenue*, Fraser of Allander Institute Research Monograph No. 8 (University of Strathclyde, 1980); UKOOA, *Submission to the Chancellor of the Exchequer on the 1981 Changes to the UK Oil Tax Regime* (UKOOA, London, October 1981); and the IFS Committee, 'The Taxation of North Sea Oil' (IFS, London, December 1981) provide some of the most vociferous comments. See also Chapter 4.

9. See Wood, Mackenzie & Company, *The North Sea in Perspective* (Wood, Mackenzie & Company, Edinburgh, April 1979).
10. A 'block' is used here and below as a unit of measurement, so that half and quarter blocks may be added.
11. A diagrammatic map of the basins can be found in the Department of Energy, *Development of the Oil and Gas Resources of the UK 1982* (HMSO, London, 1982), and a less schematic map in Department of Energy, *Development of the Oil and Gas Resources of the UK 1974* (HMSO, London, 1974). Our 'Northern North Sea Basin' includes the 'Central North Sea Basin' and the 'Central Plateau' in the diagrammatic map.
12. See C. Rowland, 'The Economics of Oil Taxation' (unpublished Ph. D. thesis), chapter 6, for an elaboration of these properties, particularly Appendix B to that chapter.
13. Department of Energy, *Press Notice Reference to No. 43* (London, 12 March 1981) describes the 'company nomination area' in the seventh round of licensing and details the four blocks allocated in this fashion at the end of 1980. The 'company nomination area' was a 'defined area of the northern North Sea from which companies were able to apply for any block or blocks of their own selection'. 'Big companies to benefit in oil licence round',*Financial Times*, 18 May 1982, notes that this practice is to be extended into the eighth round of licensing.
14. This is clealy explained in C. Robinson, 'The Future of Oil Prices', paper presented to the Operations Research Forecasting Study Group, 1982, and P. Stevens, *The World Oil Equation and the Future of Middle East Oil*, Surrey Energy Economics Discussion Paper (University of Surrey, Guildford, 1982).
15. See G. Ray and C. Rowland, *The Outlook for Oil Prices in the Medium Term,* Energy Economics Discussion Paper (University of Surrey, Guildford, January 1980,, for a discussion of some of these short-run forces.
16. See Robinson and Morgan, *North Sea Oil in the Future*, particularly Chapter 6 or C. Robinson and C. Rowland, 'An Economic Analysis of British North Sea Oil Supplies', in OECD, *Workshops on Energy Supply and Demand* (OECD, Paris, 1978), where the impact of sunk costs is explained more fully.
17. See Appendix in Ray and Rowland, *The Outlook for Oil Prices*.
18. Department of Energy, *Energy Projections 1979* (HMSO, London, 1979).
19. D. Naylor and S. N. Mounteney, *Geology of the North West European Continental Shelf*, Vol. 1 (Graham Trotman, London, 1975); R. M. Pegnum, G. Rees and D. Naylor, *Geology of the North West European Continental Shelf*, Vol. 2 (Graham Trotman, London, 1975); *The Times, The Concise Atlas of the World* (Times Books, London, 1975); Department of Energy, *Development of the Oil and Gas Resources of the UK* (HMSO, London, 1975).

Index

Note: entries in *italic type* refer to appendices.